Table of Contents

Page

List of Figures

List of Tables

AIR VEHICLE OPTIMAL TRAJECTORIES FOR
MINIMIZATION OF RADAR EXPOSURE

I. Overview

The notion of utilizing military unmanned air vehicles (UAVs) for missions other than target drones and reconnaissance is at hand. In the near future, no longer will the UAV be digitally tied to a remote pilot; instead, it will be an autonomous vehicle, making it's own in-flight decisions concerning routing, target recognition, and mission success or failure. It will be necessary for the on-board system to handle these tasks quickly and efficiently to provide for successful mission completion. This research examines the avoidance of single and multiple radars for on-board optimal trajectory planning of an autonomous air vehicle.

The best way to get from the start to the finish has always been of great interest, whether it be for a military bombing mission through enemy territory, satellite flight to orbit with minimal fuel expenditure, or the best way to ship a package from New York to Los Angeles. In the last quarter century, advances in robotics have also involved the pursuit of optimal trajectories while avoiding obstacles. A common denominator for all of these problems is the need to balance the benefits and risks involved with the different approaches. Scientists, mathematicians and engineers have devised clever ways to measure and calculate "optimal" paths from start to finish while minimizing each scenario's inherent risks. Approaches for solving this problem can be categorized into trajectory optimization, route optimization, and analogous-problem formulations.

1.1 Review of Optimal Path Planning

1.1.1 Trajectory Optimization. Trajectory optimization is a path planning approach which seeks the minimization of a specific performance index or cost function through the determination of the time history of the states and/or state control variables of a dynamic model. The calculus of variations, optimal control theory and dynamic programming are techniques commonly used to solve trajectory optimization problems. Many of the classic trajectory optimization problems have been solved using these techniques, such as the isoperimetric problem, Zermelo's navigation problem, the inverse square law describing the orbit of a particle, and many minimum time problems, e.g. light through a medium. Newton, inventor of calculus, solved a trajectory optimization problem, Bernoulli's brachistochrone, in the late 17th century [6]. This is by no means an inclusive list; see [6], [8], [10], and [20] for more examples.

Shapira [35] used optimal control theory to formulate a bang-level-bang type controller for optimal trajectories given a specified initial and final position and heading, where "bang" indicates application of a maximum rate of turn. Simplified aircraft equations of motion were used, with the performance index being a minimum time mission with constrained control usage. Analytic solutions were obtained and verified using numerical techniques.

Also attacking the trajectory optimization problem analytically, Pachter [26] found a closed-form solution to the problem of minimizing the amount of radar energy received by a radar from a constant-velocity target aircraft. His solution highlighted the result that there is a geometric limit for which an unconstrained solution exists; thus, blindly solving the problem numerically outside of this limit could lead to incorrect solutions unknowingly being labelled "optimal".

While Pachter and Shapira were successful in finding analytic solutions to specific problems, more commonly the complexity of the nonlinear optimization problems lead away from analytic solutions and toward the application of numerical methods. Bryson's text *Dynamic Optimization* [6] includes a comprehensive set of numerical algorithms and MATLAB code especially for trajectory optimization. MATLAB itself has a set of widely used optimization codes available in the Optimization Toolbox [7]. In addition, many other commercial software libraries exist such as Numerical Recipes [29] and IMSL [39]. Since many trajectory optimization problems are formulated as two-point boundary value problems, the mathematical theory and numerical methods of solution are well known (e.g. [32]) and have been implemented in many computer languages. Each of these numerical methods require problem dependent user modifications.

Psiaki and Park [30] posed the optimal control trajectory optimization problem as a time-varying LQR problem with equality constraints. By taking advantage of a unique formulation of the cost and constraint structure, the problem is solved using a hyper-cube message-passing parallel processor, an iterative technique employing variable reduction and partial solutions at each step. The results indicate that this technique has quicker solution times for large problems as compared to the matrix Riccati equation.

Vian [38] and Rao [31] used singular perturbation techniques to reduce the order of the airplane equations of motion, and developed a cost functional minimizing risk, fuel burn, and time enroute. They then applied Pontryagin's Maximum Principle, an equivalent formulation of the calculus of variations, and a Fibonacci search algorithm to numerically minimize the cost functional.

1.1.2 Route Optimization. Route optimization is the desire for an optimal set of vehicle waypoints to navigate through obstacles to meet a desired objective. These route planning problems often employ a discretization or gridding scheme for the area involved, and include boundaries on the space where waypoints can be defined. Dynamic programming or graph-based techniques are used to solve these problems numerically, utilizing efficient search methods and heuristics to ease the computational burden. Sub-optimal paths usually result from these methods due to the gridding scheme incorporated. Increasing the grid mesh will give a better solution, but is often not feasible due to the high computational cost.

Selvestrel [33] presents a method for generating optimal routes using a modified A^* heuristic search method. The A^* algorithm is an efficient graph-searching technique; the major drawback is its computational burden, as it is exponential in space requirements. Selvestrel reduces the search space using new heuristics and a technique to "prune" states from the search. His method provides an A^* solution without loss of optimality; however, the new heuristic and pruning are sensitive to problem variations and may sometimes offer little benefit.

Another common graphical method is the use of Voronoi diagram search, e.g. [18] and [23]. A Voronoi diagram is more general than this, but one application is to create the search graph for route planning. For example, given a set of known radar locations, the Voronoi diagram is constructed of polygons whose edges are equidistant from all of the neighboring radars. Travel along these edges guarantees avoidance of threats, but it may not represent the best path from start to finish. Meng [23] employed this technique for robotic air vehicles traversing through mountainous terrain, with excellent results.

1.1.3 Analogous-Problem Formulation. Analogous-problem formulation entails transforming the path planning problem to an entirely different problem which has either already been solved or has convenient method of solution.

The idea of using the physical principle of potential fields has been used to determine optimal paths for autonomous air vehicles. This method utilizes a grid with a system of connections between nodes on the grid that are based upon physical laws, such as attraction and repulsion of magnetic fields. At the Air Force Research Laboratory, several researchers have employed this technique. Bortoff [3] used this method to solve the path planning problem against radars by representing the path as a chain of masses, interconnected by springs and dampers. The radar sites generate virtual force fields proportional to the $1/R^4$ distance law, ultimately "pushing" the spring-mass chain into its potential energy minimum, a weighted sum of path length and distance from the radar. The resultant locations of the masses define the waypoints of the path, connected by straight line segments. McFarland [22] also employed the potential field approach against monostatic radars while studying the effects of minimizing radar cross section through vehicle orientation.

Pellazar [27] and Min [24] used genetic algorithms to attack the path planning problem. Genetic algorithms are a probabilistic search technique based on the principles of biological evolution, natural selection and genetic recombination. The algorithms perform an adaptive search of promising regions in the search space. The population is rated on its performance (objective function) and members (solutions) are rewarded or eliminated proportionally to their "fitness". Thus, a "survival of the fittest" scenario develops and eventually over many iterations the optimal solution is the only one left. Pellazar's results indicated that near-optimal trajectories were found, but that extensive computation time was necessary to produce desirable solutions. He notes that future parallel processing architectures will be more suited for application of this approach.

1.2 Problem Statement

1.2.1 Objectives. Optimal path planning research to date has been extensive, and many techniques exist to formulate and solve the problem. A common thread through much of the literature is that the formulations often include complicated dynamics and multiple constraints. This results in complex problems whose solution can often only be found utilizing numerical techniques. By approaching a simple problem first and understanding its solution, insights can be gained which can later be applied to more complicated problems. This research aims to explore the feasibility of geometric, deterministic solutions which, while suboptimal, approximate the optimal solution to within acceptable limits.

This research follows the work of Pachter [26], investigating the problem of minimizing the radar exposure of an air vehicle. With the existence of an analytic solution, an exploration of numerical solutions to the single radar problem can be evaluated. In addition, a similar formulation involving two radars is investigated. The specific objectives are to:

1. Apply several numeric techniques to solve the single-radar problem, evaluate the optimal path cost for several scenarios and compare the results with the analytic solution. Examine the sensitivity of the cost to changes in the path length.

2. Formulate the problem of trajectory optimization for a single vehicle against two radars.

3. Analyze the optimal paths travelling between two radars for several geometrically symmetric scenarios and compare the resultant optimal paths against the direct (i.e. straight line) path and a path following the locus of equal radar power.

1.2.2 Approach. The optimization for the single-radar scenario is performed in MATLAB using three numerical methods: a forward shooting method developed by Hebert [12] for solving two-point boundary value problems, a sequential quadratic programming (SQP) routine developed at AFIT [16], and an optimal control algorithm from Bryson [6]. The two-radar problem is evaluated utilizing the forward shooting code, and a discrete approximation to the performance index is found.

1.2.3 Scope. This research is limited to exploring the constant-velocity single vehicle scenario. It is easy enough to complicate matters by introducing additional constraints such as time, fuel burn, turn radius limits, etc. and allowing the velocity to change with time. The goal here is to understand the fundamental problem so that when additional constraints are prescribed, the insights gained from this problem can be applied and perhaps an easier and better approach to the optimal solution can be found.

II. Methodology

2.1 A Brief Review of the Calculus of Variations

The calculus of variations was developed in the 16th and 17th centuries by the great mathematicians of the time: Euler, Lagrange, Leibniz, Bernoulli and Newton. It is a branch of calculus concerned with examining extremal problems under conditions more general than the ordinary theory of maxima and minima. Specifically, it is concerned with the extremization of expressions in which entire functions must be determined. I will summarize Gelfand's formulation [10]; for the full derivation, I recommend Fox [8] and Gelfand [10] for their classical treatment of the subject.

Suppose we have the following functional

$$J[y] = \int_a^b F(x, y, y') \, dx, \quad y(a) = A, \quad y(b) = B, \tag{2.1}$$

where $J[y]$ is defined on some normed linear space. The *increment* of this functional is defined as

$$\Delta J[h] = J[y + h] - J[y], \tag{2.2}$$

where $h = h(x)$ corresponds to the increment of the "independent variable" $y = y(x)$. If y is fixed, then $\Delta J[h]$ is a functional of h. Suppose $\Delta J[h]$ is expressed as

$$\Delta J[h] = \varphi[h] + \varepsilon \parallel h \parallel, \tag{2.3}$$

where $\varphi[h]$ is a linear functional and $\varepsilon \to 0$ as $\parallel h \parallel \to 0$. $J[y]$ is then called differentiable with $\varphi[h]$ called the *variation* of $J[h]$ and is denoted by $\delta J[h]$.

Theorem 2.1.1. *A necessary condition for the differentiable functional $J[y]$ to have an extremum for $y = \hat{y}$ is that its variation vanish for $y = \hat{y}$, i.e., that*

$$\delta J[y] = 0$$

for $y = \hat{y}$ and all admissible h.

Proof. See Theorem 2, Section 3 in [10]. □

Theorem 2.1.1 leads us to the Euler equation, the foundation of modern calculus of variations.

Theorem 2.1.2. *Let $J[y]$ be a functional of the form*

$$\int_a^b F(x, y, y')\, dx$$

defined on the set of functions $y(x)$ which have continuous first derivatives in $[a, b]$ and satisfy the boundary conditions $y(a) = A$, $y(b) = B$. Then a necessary condition for $J[y]$ to have an extremum for a given function $y(x)$ is that $y(x)$ satisfy Euler's equation

$$F_y - \frac{d}{dx} F_{y'} = 0. \tag{2.4}$$

Proof. See Section 4.1 in [10]. □

In general, the result of Euler's equation will be a second order differential equation. The resulting extremals will be either minimums, maximums, or saddle points; the Euler equation does not distinguish which. For this research it will be evident if the solution is a candidate minimum, as the path will decidedly head further away from the radars. Hebert has shown that for the single radar case that the second-order necessary conditions for a minimum are met [12].

The calculus of variations theory has been extended to encompass problems with varying initial and final points, as well as interior corner conditions; that is beyond the scope of this study and interested readers are again referenced to Fox and Gelfand.

2.2 A Brief Review of Optimal Control

Optimal control is another method by which to minimize a performance index. It attempts to optimize by finding the control histories for a dynamic system for a given time period. Thus, it is an indirect method of determining the optimum; the goal is to minimize the performance index by finding the time history of the control vector $u(t)$ instead of looking for the states $x(t)$ themselves. A summary of the method will be provided here; Bryson [6] and Lewis [19] are excellent sources for the full derivation.

Consider a system described by a state vector $x(t)$. The choice of a control vector $u(t)$ will determine the rate of change of the state vector, i.e.,

$$\dot{x} = f(x, u, t).$$

(2.5)

The Bolza formulation is a common form of the performance index,

$$J = \phi[x(t_f)] + \int_{t_0}^{t_f} L(x, u, t)\, dt,$$

(2.6)

with t_0, t_f, and $x(t_0)$ specified. An equivalent formulation is called the Mayer formulation; it is formed by augmenting the state vector by one state $x_{n+1}(t)$ and defining

$$\dot{x}_{n+1} = L(x, u, t),$$

then

$$x_{n+1}(t_f) = \int_{t_0}^{t_f} L(x, u, t)\, dt + x_{n+1}(t_0),$$

and the performance index becomes

$$J = \phi[x(t_f)] + \int_{t_0}^{t_f} L(x, u, t)\, dt = \phi[x(t_f)] + x_{n+1}(t_f) - x_{n+1}(t_0) = \bar{\phi}[\bar{x}(t_f)],$$

where the states are now $\bar{x} \triangleq [x, x_{n+1}]^T$.

In many optimization problems there are constraints on the terminal point, i.e. constraint relations where the terminal state is specified,

$$\psi[x(t_f)] = 0.$$

2-3

The optimal control problem formulation in Mayer form can now be summarized as choosing $u(t)$ to minimize

$$J = \phi[x(t_f)],\tag{2.7}$$

subject to differential constraints

$$\dot{x} = f(x, u, t),\tag{2.8}$$

and boundary constraints

$$x(t_0) = x_0,\tag{2.9}$$

$$\psi[x(t_f)] = 0.\tag{2.10}$$

To find the optimal solution, the performance index (2.7) is augmented by adjoining the differential constraints (2.8) and boundary constraints (2.10) using Lagrange multipliers $\lambda(t)$ and ν,

$$\bar{J} = \phi[x(t_f)] + \nu^T \psi[x(t_f)] + \int_{t_0}^{t_f} \lambda^T(t)\{f[x, u, t] - \dot{x}\}\, dt.$$

Define the Hamiltonian to be

$$H(t) = \lambda^T(t) f[x, u, t],$$

and rewrite the augmented performance index as

$$\bar{J} = \Phi(t_f) + \int_{t_0}^{t_f} H(t)\, dt - \int_{t_0}^{t_f} \lambda^T(t)\dot{x}\, dt.$$

where $\Phi(t_f) = \phi[x(t_f)] + \nu^T \psi[x(t_f)]$. Finally, integrating the final term in $\bar{J}$ by parts gives

$$\bar{J} = \Phi(t_f) - [\lambda^T x]_{t_0}^{t_f} + \int_{t_0}^{t_f} \left[H(t) + \dot{\lambda}^T x \right]\, dt.$$

We now consider a variation in $\bar{J}$ from $u(t)$, i.e. $\delta u(t)$, which in turn will cause variations in the state histories, $\delta x(t)$,

$$\delta \bar{J} = [\Phi_x - \lambda^T]_{t=t_f} \, \delta x(t_f) + \lambda^T \, \delta x(t_0) + \int_{t_0}^{t_f} \left[(H_x + \dot{\lambda}^T) \, \delta x + H_u \, \delta u \right] dt, \qquad (2.11)$$

and will lead us to the necessary conditions for an optimal solution. If we choose $\dot{\lambda}^T$ such that the variations in δx disappear,

$$\dot{\lambda}^T = -H_x, \qquad (2.12)$$

and with boundary conditions

$$\lambda^T(t_f) = \Phi_x(t_f), \qquad (2.13)$$

then (2.11) becomes

$$\delta \bar{J} = \int_{t_0}^{t_f} H_u \, \delta u \, dt. \qquad (2.14)$$

Since $x(t_0)$ is known, $\delta x(t_0) = 0$. In order to determine a stationary point, from Theorem 2.1.1 $\delta \bar{J} = 0$ for any δu; this can only be satisfied if

$$H_u = 0. \qquad (2.15)$$

In summary, to find a control $u(t)$ that produces an extremal of the performance index J, the following set of equations must be solved:

Performance Index:	$J = \phi[x(t_f)]$	(2.7)
Differential Constraints/State Equations:	$\dot{x} = f(x, u, t)$	(2.8)
Co-State Equations:	$\dot{\lambda}^T = -H_x = -\lambda^T f_x$	(2.12)
Natural Boundary Conditions:	$x(t_0) = x_0$	(2.9)
	$\psi[x(t_f)] = 0$	(2.10)
	$\lambda^T(t_f) = \phi_x(t_f)$	(2.13)
Optimality Condition:	$H_u = \lambda^T f_u = 0$	(2.15)

The state and co-state equations, (2.8) and (2.12), constitute a set of coupled differential equations. They define a two-point boundary value problem, since the boundary conditions required for solution are the initial state, equation (2.9), and the final state and costate, equations (2.10) and (2.13).

2.3 A Review of Minimizing Radar Exposure in Air Vehicle Path Planning

This section summarizes Pachter's solution [26], as well as some alternate performance index formulations. A discrete formulation of the performance index is presented, as it provides a way to utilize numerical methods such as SQP to solve the problem. The full application of the Euler equation is addressed; the resulting second-order differential equation is used with a two-point boundary problem solver as an alternate method of numerical solution.

2.3.1 Continuous Performance Index.

The amount of power received by a radar is given by the radar range equation [37]

$$P_r = \frac{P_t G A_e \sigma}{(4\pi)^2 R^4},$$

where P_t is the power of the radar transmitter, G is the transmitting gain, A_e is the effective area of the receiving antenna, σ is the radar cross section of the target, and R is the distance of the target to the radar. For the purposes of this study, the power received by the radar is considered only a function of range to the target, i.e.

$$P_r \propto \frac{1}{R^4}.$$
(2.16)

Given a radar located at the origin and the geometry in Figure 2.1, Pachter [26] showed that an objective function for minimizing P_r is

$$J = \int_0^{\frac{\ell}{v}} \frac{1}{R^4(t)}\, dt,$$
(2.17)

where v is the (constant) speed of the aircraft and ℓ is the path length. Rewriting equation (2.17) in polar coordinates yields

$$J[R(\theta)] = \int_{\theta_0}^{\theta_f} \frac{\sqrt{\dot{R}^2 + R^2}}{R^4}\, d\theta,$$
(2.18)

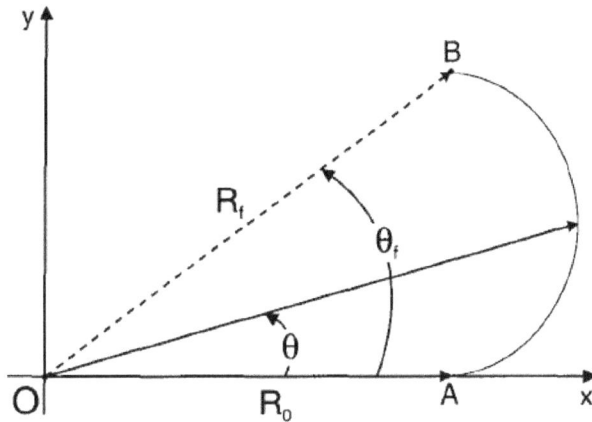

Figure 2.1   One Radar Problem Geometry

with boundary conditions

$$R(0) = R_0, \tag{2.19}$$

$$R(\theta_f) = R_f, \quad 0 < \theta \le \theta_f. \tag{2.20}$$

2.3.2 Problem Solution. The geometry of the problem is shown in Figure 2.1. The problem was posed as a problem in the calculus of variations. Pachter's solution is stated in Theorem 2.3.1.

Theorem 2.3.1 (Pachter-Hebert Theorem). *The optimal trajectory which connects points A and B at a distance R_o and R_f from the radar located at the origin O, where θ_f is the angle $\angle AOB$, and minimizes the exposure to the radar according to Eqs. (2.18)-(2.20), is*

$$R^*(\theta) = R_o \sqrt[3]{\frac{\sin(3\theta + \phi)}{\sin\phi}}, \quad 0 < \theta \le \theta_f \tag{2.21}$$

where

$$\phi = \text{Arctan}\left(\frac{\sin 3\theta_f}{\left(\frac{R_f}{R_o}\right)^3 - \cos 3\theta_f}\right). \tag{2.22}$$

Moreover, the length of the optimal path is given by the integral

$$l^* = \frac{R_o}{\sqrt[3]{\sin\phi}} \int_0^{\theta_f} [\sin(3\theta + \phi)]^{-\frac{2}{3}} \, d\theta, \tag{2.23}$$

and the cost function explicitly evaluates to

$$J^* = \frac{1}{3R_o{}^3} \frac{\sin 3\theta_f}{\sin(3\theta_f + \phi)}. \tag{2.24}$$

This result holds provided $0 < \theta_f < \frac{\pi}{3}$. However, if $\frac{\pi}{3} \le \theta_f \le \pi$, then an optimal path does not exist and a constraint on the path length, l, must be included to render the optimization problem well posed.

Proof. See Theorem 1 in [26]. □

The solution for the optimal path, equation 2.21, as well as the optimal cost J^*, equation 2.24, are extremely useful; they will be the baseline for which the numerical methods of Chapter 3 are referenced.

2.3.3 Discrete Approximation of the Performance Index. In order to apply a numerical parameter optimization method such as SQP, it is necessary to approximate the performance index for an optimal trajectory. Pachter considers the optimal path from points A to B as a series of waypoints connected by straight line segments. Each line segment can be written as a line in polar coordinates as a function of θ,

$$R(\theta) = \frac{r_1 r_2 \sin(\theta_2 - \theta_1)}{r_1 \sin(\theta - \theta_1) - r_2 \sin(\theta - \theta_2)},$$ (2.25)

with the derivative

$$\dot{R}(\theta) = r_1 r_2 \sin(\theta_2 - \theta_1) \frac{r_1 \cos(\theta - \theta_1) - r_2 \cos(\theta - \theta_2)}{[r_1 \sin(\theta - \theta_1) - r_2 \sin(\theta - \theta_2)]^2}.$$ (2.26)

Substituting these expressions into the continuous performance index, equation (2.18), and integrating from the point (r_1, θ_1) to the next point (r_2, θ_2) yields

$$
\begin{aligned}
J_{1 \to 2} &= \int_{\theta_1}^{\theta_2} \frac{\sqrt{\dot{R}^2 + R^2}}{R^4} \, d\theta \\
&= \frac{\sqrt{r_1^2 + r_2^2 - 2 r_1 r_2 \cos \Delta\theta}}{4(r_1 r_2 \sin \Delta\theta)^3} [4 r_1 r_2 \Delta\theta (\cos \Delta\theta - sin\Delta\theta) - (r_1^2 + r_2^2)(2\Delta\theta - \sin 2\Delta\theta)],
\end{aligned}
$$ (2.27)

where $\Delta\theta = \theta_1 - \theta_2$. Thus the continuous dependence upon θ has been eliminated and the cost can be determined for any given pair of points (r_1, θ_1) and (r_2, θ_2). The total cost for a path of N line segments is simply

$$\tilde{J}^* = \sum_{i=1}^{N} J_{i \to i+1}.$$ (2.28)

Provided $\Delta\theta$ is chosen sufficiently small, this gives an accurate presentation of the continuous performance index, allowing the variational problem to be solved as a parameter optimization problem using SQP.

2.3.4 Application of the Euler Equation.   The following application of the full Euler equation defines the second-order differential equations describing the extremal path, which are easily converted to first-order form for use in numerical integrators.

Given the cost functional for flight against one radar,

$$J[R(\theta)] = \int_{\theta_0}^{\theta_f} \frac{\sqrt{\dot{R}^2 + R^2}}{R^4} \, d\theta \,, \tag{2.18}$$

apply the Euler equation to find the differential equation describing the optimal path,

$$\frac{\partial L}{\partial R} - \frac{d}{d\theta}\left(\frac{\partial L}{\partial \dot{R}}\right) = 0 \,, \tag{2.4}$$

$$L = \frac{\sqrt{\dot{R}^2 + R^2}}{R^4} \,,$$

$$\frac{\partial L}{\partial R} = \frac{1}{R^3 \sqrt{R^2 + \dot{R}^2}} - \frac{4\sqrt{R^2 + \dot{R}^2}}{R^5} \,,$$

$$\frac{\partial L}{\partial \dot{R}} = \frac{\dot{R}}{R^4 \sqrt{R^2 + \dot{R}^2}} \,,$$

$$\frac{d}{d\theta}\left(\frac{\partial L}{\partial \dot{R}}\right) = \frac{-4\dot{R}^2}{R^5 \sqrt{R^2 + \dot{R}^2}} + \frac{\ddot{R}}{R^4 \sqrt{R^2 + \dot{R}^2}} - \frac{\dot{R}(R\dot{R} + \dot{R}\ddot{R})}{R^4(R^2 + \dot{R}^2)^{3/2}} \,.$$

Simplifying and solving for $\ddot{R}$ results in

$$\ddot{R} = -3R - \frac{2\dot{R}^2}{R} \,, \tag{2.29}$$

2-10

with the boundary conditions

$$R(0) = R_0, \qquad (2.19)$$

$$R(\theta_f) = R_f, \quad 0 < \theta \le \theta_f. \qquad (2.20)$$

The conversion to first order form for use in MATLAB is as follows,

$$x_1 = R, \qquad (2.30)$$

$$x_2 = \dot{R}, \qquad (2.31)$$

$$\dot{x}_1 = x_2, \qquad (2.32)$$

$$\dot{x}_2 = \ddot{R} = -3\,x_1 - \frac{2\,x_2^2}{x_1}, \qquad (2.33)$$

with boundary conditions

$$x_1(0) = R_0, \qquad (2.34)$$

$$x_1(\theta_f) = R_f. \qquad (2.35)$$

The optimal path must satisfy equations (2.30)-(2.35).

III. Trajectory Optimization Against One Radar

3.1 Overview

Variational problems often do not have an analytic solution. The non-linear differential equations that result from the Euler equation are often too difficult to solve analytically and require the application of a numerical method. Pachter's analytic solution for trajectory optimization against one radar, Theorem 2.3.1, provides a unique opportunity to contrast different numerical solution methods with a known optimal solution. This solution will be the benchmark to judge different numerical methods used to solve for the optimal trajectory.

There are many optimization techniques available to solve these problems. For this study, I applied three numerical methods: a two-point boundary value problem solver, an optimal control algorithm, and a SQP solver. All numerical optimizations were done using MATLAB, a widely used mathematics software suite.

The two-point boundary value problem solver used was developed by Hebert [12] using core MATLAB commands. It is a forward shooting method for solving differential equations. Starting with a guess for the initial conditions, the Euler equation is integrated forward, and the error between the desired final point and the calculated final point is calculated. A nonlinear root-finding algorithm is then used to drive the error to an arbitrarily close user-defined tolerance.

In *Dynamic Optimization* [6], Bryson provides several different algorithms for the solution of optimal trajectory problems, as well as the supporting MATLAB codes implementing these algorithms. The Functional Optimization with Constraints (FOPC) algorithm was used for the optimal control solution. The state equations, equations (3.7) and (3.8), are integrated forward from the specified initial conditions with an initial guess of the control vector, $u(\theta)$. The co-state equations, equations (3.9) and (3.10), are then integrated backward to determine a gradient sequence, which is used to make small changes in the control sequence, moving the solution closer to the desired final conditions. The process is repeated until the final conditions and gradient sequence are within an arbitrarily close user-defined tolerance, giving the optimal solution.

An AFIT-developed SQP solver [16] was the final numerical method evaluated. Sequential quadratic programming is a nonlinear programming technique using line search methods to systematically march towards an optimal solution. The discrete performance index, equations (2.27) and (2.28), provides the cost to be minimized for this solution technique.

The continuous and discrete performance indices for trajectory optimization against one radar were developed by Pachter [26]. These equations provide the functions necessary for use of the shooting method and SQP. A general development of optimal control was reviewed in Chapter 2; the following section determines the specific equations necessary for use of Bryson's FOPC algorithm.

3.2   Optimal Control Formulation

The one radar problem is now formulated as an optimal control problem in Mayer form. The performance index to be minimized is equation (2.18), with θ as the independent variable. To convert to Mayer form, the states are defined as $x = [R, S]^T$ and control $u = \dot{R}$. The performance index is now

$$J = S(\theta_f), \tag{3.1}$$

subject to boundary conditions

$$R(\theta_0) = R_0, \tag{3.2}$$

$$\psi \equiv R(\theta_f) - R_F = 0, \tag{3.3}$$

and differential constraints

$$\dot{R} = u, \tag{3.4}$$

$$\dot{S} = \frac{\sqrt{R^2 + u^2}}{R^4}. \tag{3.5}$$

The goal of the optimal control problem is to find $u(\theta)$ to minimize the cost function, equation (3.1), subject to the constraints (3.4)-(3.5). Adjoining the constraints to the performance index with Lagrange multipliers ν and $\lambda(\theta)$

$$J' = \phi + \nu^T \psi + \int_0^{\theta_f} (H - \lambda^T(\theta)\dot{x})\, d\theta \,, \qquad (3.6)$$

where

$$\phi = S(\theta_f) \,,$$

$$\psi = R(\theta_f) - R_F \,,$$

$$H = \lambda_R u + \lambda_S \frac{\sqrt{R^2 + u^2}}{R^4} \,.$$

In summary, the optimal control formulation starts with the state equations (2.8)

$$\dot{R} = u \,, \qquad (3.7)$$

$$\dot{S} = \frac{\sqrt{R^2 + u^2}}{R^4} \,, \qquad (3.8)$$

and co-state equations (2.12)

$$\dot{\lambda}_R = \lambda_S \left[\frac{4\sqrt{R^2 + u^2}}{R^5} + \frac{1}{R^3\sqrt{R^2 + u^2}} \right] \,, \qquad (3.9)$$

$$\dot{\lambda}_S = 0 \,, \qquad (3.10)$$

with natural boundary conditions (2.13)

$$R(\theta_0) = R_0 \,, \qquad (3.2)$$

$$\psi \equiv R(\theta_f) - R_F = 0 \,, \qquad (3.3)$$

$$\lambda_{r(\theta_f)} = \nu_r \,, \qquad (3.11)$$

$$\lambda_{s(\theta_f)} = 1 \,, \qquad (3.12)$$

and finally, the optimality condition (2.15)

$$H_u = \lambda_R + \lambda_S \, \frac{u}{R^4 \sqrt{R^2 + u^2}} = 0 \, . \tag{3.13}$$

These equations constitute the optimal control formulation used for analyzing flight against one radar.

3.3 Optimization Results

Four scenarios were examined using each of the numerical techniques, as well as the analytic solution, Table 3.1. The radar is always located at the origin, and the starting endpoint was always at $R_0 = 1$, $\theta_0 = 0°$, so θ_f indicates the angle traversed. Each path consists of one hundred points, linearly spaced in θ.

Case 1 is a fairly simple problem, a constant radius at the endpoints and a shallow θ_f. Case 2 kept the shallow angle, but extended the endpoints by an order of magnitude. Case 3 is similar to Case 1, but θ_f approaches the 60° limit for an analytic solution. Case 4 approaches the extremes in both endpoint separation and θ_f. The goal was to examine numerically diverse problems and explore the extremities of endpoint position and change in angle.

The cost and path length of the optimal trajectory was tabulated for each case. The optimal control solver and SQP each minimize the objective cost during the search for an optimal trajectory, as per their formulation, so the cost was immediately available. The shooting method does not minimize the cost, but directly solves the Euler equation, so the discrete cost function was used to calculate the cost for this method. The cost of

Table 3.1   Scenarios Evaluated for Flight Against One Radar

Case	R_0	R_f	θ_f
1	1	1	20°
2	1	10	20°
3	1	1	59°
4	1	10	59°

the shortest path length trajectory, i.e. a straight line from R_0 to R_f, is also included as another comparison.

Each trajectory solution was evaluated by the discrete cost function and the FOPC algorithm. To use the FOPC algorithm, the control $u(\theta)$, the gradient of the path, was numerically evaluated by using a forward difference at the left end, a backward difference at the right end, and central difference in the interior. This was input as the initial condition to the FOPC algorithm and was then run for one iteration to determine the trajectory cost. Additionally, the path length for each solution was calculated using a summation of the distances between consecutive points, and was added as a constraint for two of the cases. The path length is an important factor to consider, as it defines the amount of time the air vehicle will be in the radar's coverage; the shorter the path length, the less exposure the vehicle will have to the radar. Also, in a realistic scenario, the path length would ultimately be constrained by fuel considerations. As was stated in Chapter 1, this research is concerned with understanding the simplest problem first; additional constraints can be easily added later when a more complex problem is desired. That being said, the addition of the path length constraint provides some added insight into the nature of the performance index of this problem.

A couple of notes should be made regarding the operation of the numerical solvers. The shooting method requires an initial guess of the derivative, and it is often necessary to iterate on this guess to converge to a solution. This is a simple process, and the solver either fails or finds a solution fairly quickly. The SQP method requires a full initial path. During the calculation of the cost function, the path is varied slightly in the search for the optimal path. This makes the SQP run time quite a bit longer than the shooting method since more calculations are required. The FOPC algorithm requires a full initial control, which is used to integrate the state equations forward, and then the co-state equations are integrated backward. Thus, for each iteration, at least two integrations are being performed, which increases run time. Additionally, the formulation involves a step-size parameter, k, and a constant η, which is the desired change in terminal constraints for the following iteration. Each of these are user inputs and their settings directly affect the convergence of the algorithm. Bryson himself says that it takes "a little practice" [6] to learn how to vary k

so that an effective step-size is reached. The process of varying these parameters so that the problem converges to a solution is very tedious, frustrating and non-intuitive. While the algorithm appears to be an efficient way of solving the problem, it is the most difficult to use, and I believe that the manual selection of these parameters definitely affected the outcome of the solver and greatly increased the time to find a solution.

  3.3.1 Case 1. This case examines a fairly short, small $\Delta\theta$ trajectory. The results are shown in Table 3.2 and Figure 3.1. The numerical methods had little trouble finding the optimal path for this scenario. The objective cost and path lengths are nearly identical for all of the solutions. Note that while the length of the straight line path is slightly less than the others, the cost is higher; this makes sense, since the direct path stays closer to the radar throughout the trajectory. This finding will be true for each of the scenarios investigated.

Table 3.2 Optimal Cost, J^*, and Path Length, ℓ^*, for Case 1: $R_0 = 1$, $R_f = 1$, $\theta_f = 20\,^\circ$

| | | | J^* *Evaluated by* | |
Path From	ℓ^*	Analytic	Discrete	FOPC
Analytic	0.378020	0.333333	0.333336	0.333340
Line	0.347296	–	0.361783	0.361777
Shooting	0.378020	–	0.333336	0.333340
SQP	0.378013	–	0.333336	0.333340
FOPC	0.378022	–	0.333336	0.333333

  3.3.2 Case 2. For this scenario, R_f was extended from 1 to 10 while keeping the angle sweep at $\theta_f = 20\,^\circ$. The optimization results are presented in Table 3.3 and Figure 3.2. This proved to be a slightly more difficult problem for some of the numerical methods to optimize. The shooting method found the analytic optimum and matched the cost function when evaluated by both the discrete and FOPC algorithms. The SQP formulation approached the optimal path as well, only differing from the optimal cost by 1.2×10^{-5} and 2.2×10^{-5} for the discrete and FOPC evaluations, respectively, while achieving a shorter path length.

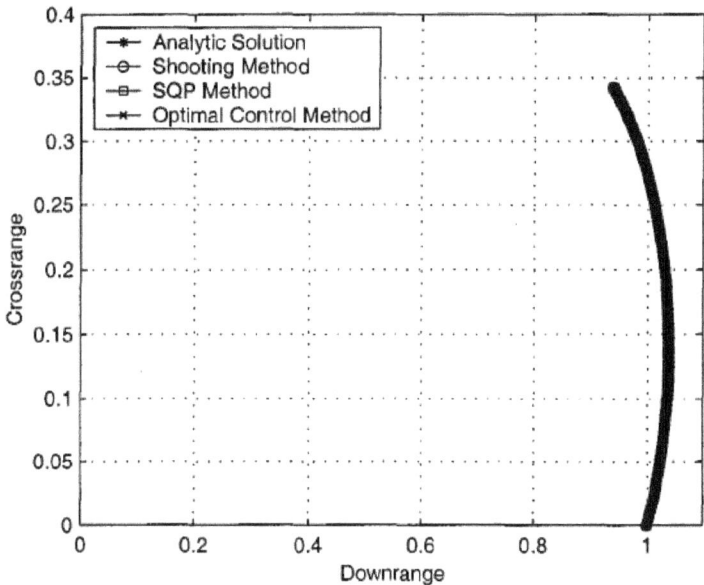

Figure 3.1   Optimal Trajectories for Case 1: $R_0 = 1$, $R_f = 1$, $\theta_f = 20°$

Interestingly, the FOPC algorithm found a minimum that didn't have an excessively high cost relative to the optimal solution, but appears to have a discontinuity in the calculation from the initial point to the first waypoint. Further attempts at determining a solution resulted in the same path. The same behavior is witnessed in Case 4, in which the endpoints are again separated by a factor of 10, but does not occur in Cases 1 and 3 when $R_f = R_0$. It appears that the endpoint geometry of Cases 2 and 4 create numerical difficulties for the FOPC algorithm. The control $u = \dot{R}$ is numerically calculated in the shooting method as well as FOPC; this allows a comparison of the initial values of $\dot{R}(\theta)$, Table 3.4. In Cases 2 and 4, the path is changing quite rapidly with respect to θ at the beginning of the path. It is believed that this, combined with the aforementioned

Table 3.3   Optimal Cost, J^*, and Path Length, ℓ^*, for Case 2: $R_0 = 1$, $R_f = 10$, $\theta_f = 20°$

		J^* Evaluated by		
Path From	ℓ^*	Analytic	Discrete	FOPC
Analytic	9.850997	0.333167	0.333169	0.333179
Line	9.066761	–	0.348426	0.348447
Shooting	9.850991	–	0.333169	0.333179
SQP	9.806455	–	0.333179	0.333189
FOPC	9.935508	–	0.333235	0.333234

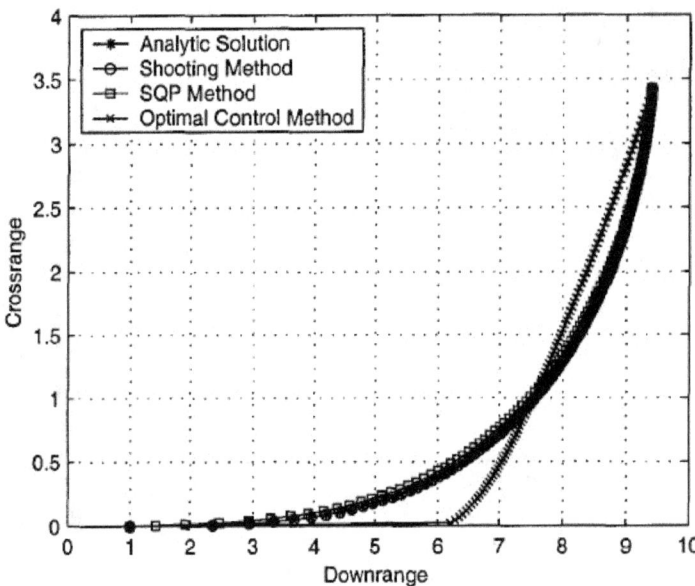

Figure 3.2   Optimal Trajectories for Case 2: $R_0 = 1$, $R_f = 10$, $\theta_f = 20°$

difficulties in choosing the step-size parameter, k, is the source of this problem for the FOPC algorithm.

Noting the reasonable cost from the FOPC optimization paired with an unusual path, a path length constraint was added to the SQP formulation to explore the relationship between path length and objective cost, Figure 3.3. The path length was iteratively constrained from the length of the direct path to the length of the analytic solution and plotted versus path length. It is interesting to note that the path length needs to change only a very small amount from a straight line before it is essentially constant at the optimal cost. This is because the optimal path is just slightly larger than the direct path; for different geometries, the optimal path will be much more pronounced and the Pareto curve

Table 3.4   Comparison of Initial $\dot{R}(\theta)$ for Cases 1-4

| | $\dot{R}(\theta)$ From | |
Case	Shooting	FOPC
1	0.577350	0.577362
2	1154.077988	2929.756315
3	38.182107	26.537612
4	19122.226062	7280.038907

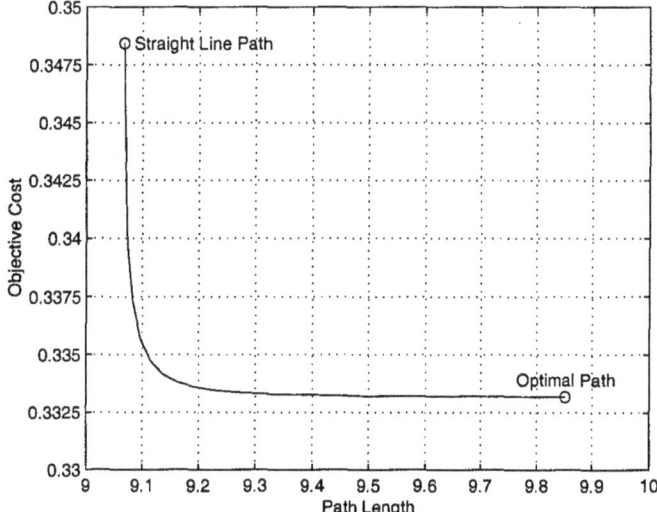

Figure 3.3   Constrained Path Length versus Cost for Case 2: $R_0 = 1$, $R_f = 10$, $\theta_f = 20°$

generated will cover a greater range of path lengths (e.g. Figure 3.6). This result is of great practical importance, because it indicates these suboptimal trajectories will provide shorter flight times with little degradation in minimizing radar exposure.

3.3.3 Case 3.   Cases 3 and 4 explore the effects of approaching the theoretical limit of solution for this problem. Pachter [26] states that the path length will approach infinity as θ_f approaches 60°, and the optimal cost will approach

$$\lim_{\theta_f \to \frac{\pi}{3}} J^* = \frac{1}{3} \left(\frac{1}{R_0^3} + \frac{1}{R_f^3} \right) , \qquad (3.14)$$

which for Case 3, $J^* \to 0.666667$, and for Case 4, $J^* \to 0.333667$. This scenario explores how well the numerical methods work when approaching this limit; Case 4 adds a large separation of endpoints to the mix.

Figure 3.4 and Table 3.5 summarize the results of Case 3. Once again, the shooting method found a solution which is nearly identical to the analytic optimal solution. It appears that the shooting method found an optimal trajectory that is shorter than the analytic solution; in reality, since both path lengths are calculated by discretizing the trajectory over only 100 points, small changes in a point's position may force one path to

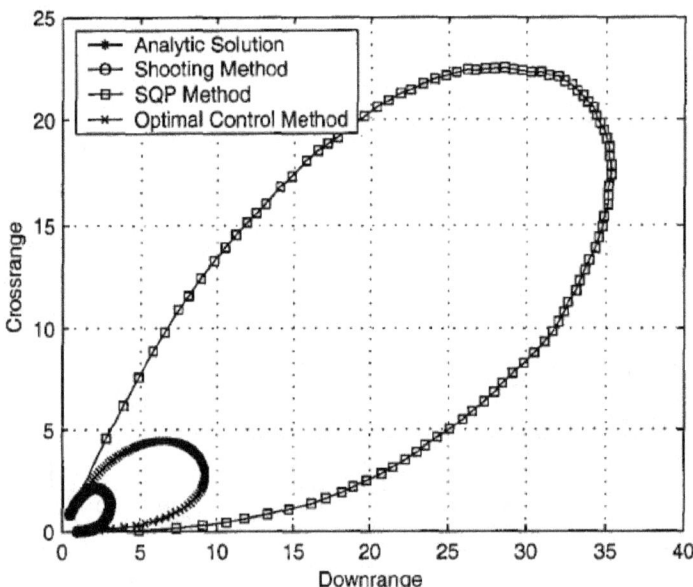

Figure 3.4   Optimal Trajectories for Case 3: $R_0 = 1$, $R_f = 1$, $\theta_f = 59°$

be smaller than the optimal. As the number of points is increased, however, both solutions approach the analytic solution's optimal length.

This does not explain, however, why the FOPC algorithm calculates a lower cost for the path generated by the shooting method and SQP than for the analytic trajectory cost. The FOPC method calculated costs less than the analytic solution for all of the methods, including the analytic solution. Also, the path generated by FOPC is larger than the analytic solution. This indicates some sort of numerical problem is occurring. The algorithm consistently gave longer path lengths; even giving the derivative of the analytically exact initial condition results in a path length much greater than that of the analytic trajectory.

Table 3.5   Optimal Cost, J^*, and Path Length, ℓ^*, for Case 3: $R_0 = 1$, $R_f = 1$, $\theta_f = 59°$

| | | | J^* Evaluated by | |
Path From	ℓ^*	Analytic	Discrete	FOPC
Analytic	6.178930	0.666438	0.666486	0.666046
Line	0.984847	–	1.430972	1.430835
Shooting	6.178541	–	0.666486	0.666016
SQP	88.252304	–	0.666704	0.649863
FOPC	20.035219	–	0.668085	0.668060

The SQP formulation also had difficulties finding the analytic optimal trajectory. The SQP solution path is nearly fifteen times longer than the analytic solution's path; the cost, however, is only 2×10^{-4} greater than the analytic cost, calculated using the discrete cost function. This makes sense; flying a longer path further from the radar will cause a minimal rise in cost due to the longer amount of exposure. Viewed in the context of the Pareto diagram of Case 2, Figure 3.3, this means that the curve extending past the optimal solution has a very shallow positive slope, which is expected.

I believe that the problems encountered by the SQP and FOPC numerical optimizers are due to the fundamental nature of the problem, i.e., the fact that as $\theta_f \to 60°$, $R^* \to \infty$. Additionally, when the function describing the relation of path length to cost is nearly constant, as is the case in the region around the path length of the analytic trajectory, it becomes increasingly difficult for the numerical solvers to distinguish what the true optimal solution is. This reinforces the statement that understanding the fundamental problem is crucial; without the knowledge of the theoretical limits, the excessive path lengths might have been considered the "optimal" solution and been no cause for further investigation. It also hints that suboptimal methods for path generation may not result in a significantly higher exposure cost. This is important as approximate methods for on-line implementation are considered.

3.3.4 Case 4. Case 4 explores the extremes in both endpoint separation and angle traversed and in doing so experiences all of the problems previously encountered in SQP and FOPC. Figure 3.5 is a plot of the optimal trajectories for each of the solvers. The SQP trajectory shoots out to an enormous path length solution and is not fully shown on the plot. While the path length is large, the cost of the SQP solution is within 1×10^{-6} of the optimal cost, Table 3.6. FOPC does not find a large path length trajectory as it did in Case 3, but it encounters problems similar to Case 2 with a quickly increasing derivative. FOPC finds a cost for the SQP solution which is less than the analytic optimum, similar to Case 3; this indicates a difficulty of the algorithm to calculate an accurate cost for extreme path lengths.

Table 3.6   Optimal Cost, J^*, and Path Length, ℓ^*, for Case 4: $R_0 = 1$, $R_f = 10$, $\theta_f = 59°$

| Path From | ℓ^* | J^* Evaluated by | | |
		Analytic	Discrete	FOPC
Analytic	53.943714	0.333666	0.333680	0.333685
Line	9.523615	--	0.498655	0.498611
Shooting	53.940363	--	0.333680	0.333685
SQP	47652.155428	--	0.333681	0.333448
FOPC	76.737776	--	0.333709	0.333698

One constant throughout this analysis is the performance of the shooting method. The solutions found by the shooting method match the analytic solution for all of the cases. Perhaps the performance of the shooting method results from directly integrating the Euler equation as opposed to the other methods. SQP calculates an approximation to the value of the cost function. It is an indirect way of satisfying the Euler equation. The optimal control equations coupled together form the Euler equation; perhaps errors are induced in the solution by the forward and backward integrations of the FOPC algorithm. Regardless, the shooting method performed exceptionally well.

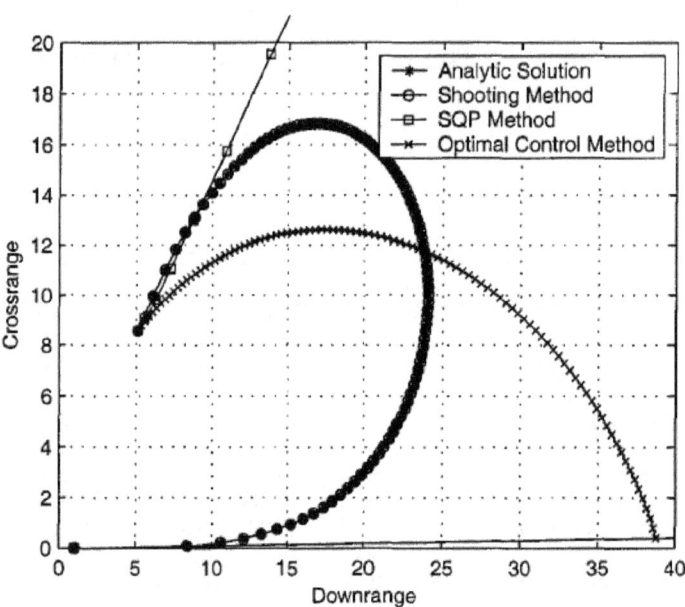

Figure 3.5   Detail of Optimal Trajectories for Case 4: $R_0 = 1$, $R_f = 10$, $\theta_f = 59°$

For this case an evaluation of path length versus objective cost similar to Case 2 was performed and is shown in Figure 3.6. The Pareto curve encompasses path lengths from the straight line path to the optimal path. From the curve it is evident that once the path length is at about 35% of the optimal path length, the cost is essentially constant. This information, taken in concert with the results from the optimizations, clearly show that suboptimal paths provide effectively the same quality of radar exposure minimization with the added benefit of shorter path lengths and therefore shorter flight times.

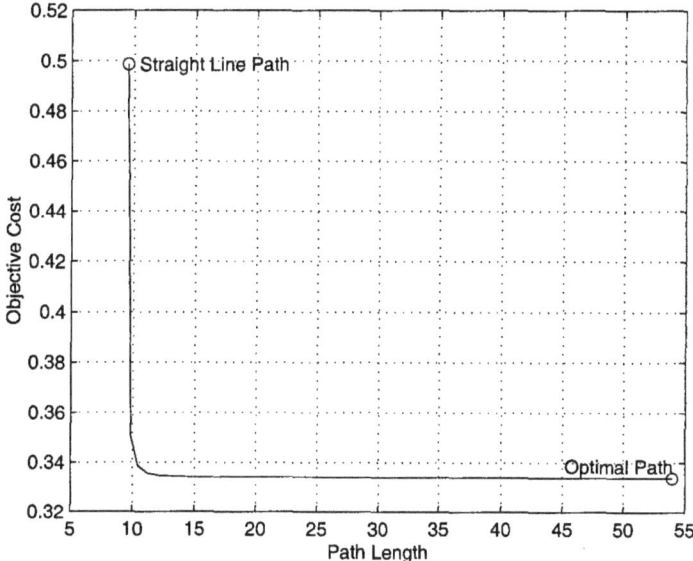

Figure 3.6   Constrained Path Length versus Cost for Case 4: $R_0 = 1$, $R_f = 10$, $\theta_f = 59°$

IV. Trajectory Optimization Against Two Radars

4.1 Overview

A natural extension to the problem of radar exposure minimization is to formulate the problem for multiple radars. This research begins this process by examining the two-radar problem. In order to solve the problem, the performance index must first be derived. The continuous and discrete performance indices are formulated, and the Euler equation is applied to determine the necessary condition for an extremal solution. While an analytic solution to the problem is desired, the complexity of the Euler equation precludes its solution; thus, the shooting method from Chapter 3 is employed to numerically solve for the optimal trajectories for several geometrically symmetric scenarios. One should note that the integrator used in the solver requires a monotonically increasing dependent variable, x. This effectively limits the path to travel between the radars; if the path were to attempt to go around one radar, it would invariably want to circle around, similar to the results in Chapter 3 (e.g. Figure 3.4), and $y(x)$ would have multiple values for a single x. Examining trajectories that travel around the radar locations is a future topic of research and will not be addressed in this study.

The radar and endpoint geometry for the scenarios examined is shown in Figure 4.1. Three parameters of the geometry were varied to examine the effects upon the optimal trajectory: the downrange distance between the radar locations (A), the crossrange distance

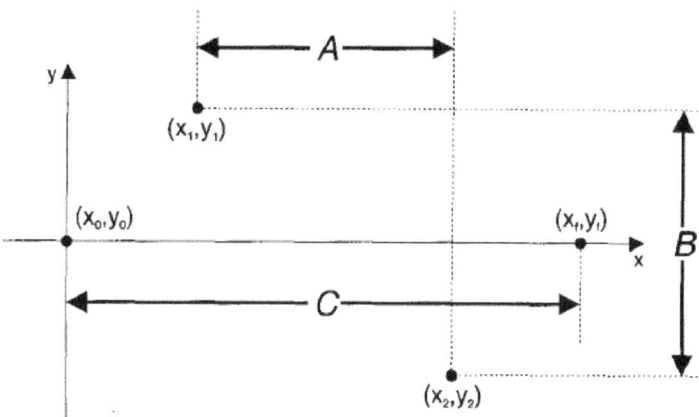

Figure 4.1   Geometry for the Two Radar Scenarios

between the radar locations (B), and the downrange distance between the initial and final point (C). When varying A and B, the endpoints of the path were fixed at $(x_0, y_0) = (0, 0)$ and $(x_f, y_f) = (1, 0)$; when varying C, the radars were fixed at $(x_1, y_1) = (0.4, 0.5)$ and $(x_2, y_2) = (0.6, -0.5)$. Two ratios of radar transmission power were examined for each of the cases: $\alpha_1/\alpha_2 = 1/1$ and $\alpha_1/\alpha_2 = 2/1$. The ordinary or weighted Voronoi trajectory, as applicable, was computed to be compared with the optimal trajectory. Table 4.1 summarizes the scenarios examined.

A common graphical technique for optimal path planning against multiple radars is to make use of the Voronoi diagram. Starting with full knowledge of the radar locations, the Voronoi diagram is constructed of polygons whose edges are equidistant from all of the neighboring radars. Hence, travel along the Voronoi edge ensures that an equal amount of power is reflected to each radar. This is true, however, only for the case where the transmission power of the radars are equal. When the radars have differing transmission powers, i.e. $\alpha_1 \neq \alpha_2$ in equation (4.1), the Voronoi edge is no longer a line but a circular arc, known as the circles of Apollonius. Apollonius of Perga, the Great Geometer, proved in his books *Conics* [1] that the locus of points whose distance from a fixed point is a multiple of its distance from another fixed point is a circle. It is easy to see that if the points have equal weight, the resulting locus is a circle of infinite radius, or a line. Since this is a widely used path planning technique, it provides a useful comparison to the path length and the objective cost of the calculated optimal trajectories. Section 4.5.1 describes the ordinary Voronoi path when $\alpha_1 = \alpha_2$, and Section 4.6.1 develops the weighted Voronoi path for $\alpha_1 \neq \alpha_2$.

Table 4.1   Scenarios for Trajectory Optimization Against Two Radars

| | | | | Varied Parameter | |
Scenario	*A*	*B*	*C*	Range	Δ
1	Varied	Fixed	Fixed	0.0- 1.0	0.2
2	Fixed	Varied	Fixed	0.4- 2.0	0.2
3	Fixed	Fixed	Varied	0.2-10.0	0.2

For the case of single vehicle flight against two radars, the power received by the radar is now considered a function of each radar's transmission power and range,

$$P_r \propto \frac{P_t}{R^4} . \tag{4.1}$$

The geometry of the problem is shown in Figure 4.2. Cartesian coordinates will be used for the formulation of flight against two threat radars as they yield a simpler performance index. First, the amount of energy received by the second radar is appended to equation (2.17)

$$J = \int_0^{\frac{\ell}{v}} \left[\frac{\alpha_1}{R_1^4(t)} + \frac{\alpha_2}{R_2^4(t)} \right] dt , \tag{4.2}$$

where α_1 and α_2 signify the transmission power of each individual radar. Consider a transformation of the integral from the time domain to the Cartesian frame with the path defined as $f = f(x, y(x))$. For convenience, y will always be considered a function of x, i.e $y = y(x)$. Now, $v = \frac{ds}{dt}$ or $dt = \frac{ds}{v}$, and ds, the element of arc length, is given in Cartesian coordinates by

$$ds = \sqrt{1 + \left(\frac{dy}{dx} \right)^2} \, dx = \sqrt{1 + \dot{y}^2} \, dx .$$

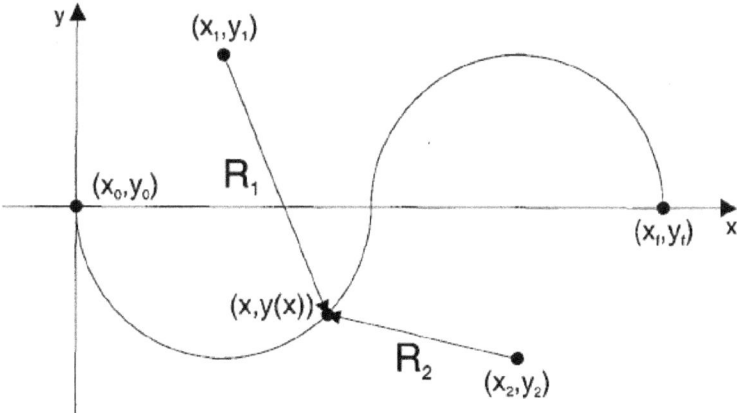

Figure 4.2   Two Radar Problem Geometry

Noting that the distance from each radar to some point on the path (x, y) is simply,

$$R_1(x, y) = \sqrt{(x - x_1)^2 + (y - y_1)^2}, \tag{4.3}$$

$$R_2(x, y) = \sqrt{(x - x_2)^2 + (y - y_2)^2}, \tag{4.4}$$

where (x_1, y_1) and (x_2, y_2) are the known radar locations in Cartesian coordinates. Equation (4.2) can now be transformed into Cartesian coordinates, yielding the performance index

$$J = \int_{x_0}^{x_f} \left(\frac{\alpha_1}{R_1^4(x, y)} + \frac{\alpha_2}{R_2^4(x, y)} \right) \sqrt{1 + \dot{y}^2} \, dx, \tag{4.5}$$

with $R_1(x, y)$ and $R_2(x, y)$ defined above. The boundary conditions are the vehicle's given initial point (x_0, y_0) and final point (x_f, y_f).

4.3 Discrete Approximation of the Performance Index

As was the case with the single radar problem, for application of nonlinear programming techniques such as SQP to the variational problem, a discrete approximation of the performance index is desired. For the two radar problem, the performance index is given in Cartesian coordinates; thus, the equation of the line segment between two points (x_j, y_j) and (x_{j+1}, y_{j+1}) and its derivative are easily determined from the point-slope form of a line,

$$y = \left(\frac{y_{j+1} - y_j}{x_{j+1} - x_j} \right) x + \frac{x_{j+1} y_j - x_j y_{j+1}}{x_{j+1} - x_j},$$

$$\dot{y} = \frac{y_{j+1} - y_j}{x_{j+1} - x_j}.$$

Substituting these results into equations (4.3)-(4.5) gives

$$J_{1 \to 2} = \int_{x_j}^{x_{j+1}} \left[\frac{\alpha_1}{R_1^4(x_j, y_j)} + \frac{\alpha_2}{R_2^4(x_{j+1}, y_{j+1})} \right] \sqrt{1 + \left(\frac{y_{j+1} - y_j}{x_{j+1} - x_j} \right)^2} \, dx, \tag{4.6}$$

with

$$R_1(x_j, y_j) = \sqrt{(x - x_1)^2 + (\frac{y_{j+1} - y_j}{x_{j+1} - x_j} x + \frac{x_{j+1} y_j - x_j y_{j+1}}{x_{j+1} - x_j} - y_1)^2}, \qquad (4.7)$$

$$R_2(x_{j+1}, y_{j+1}) = \sqrt{(x - x_2)^2 + (\frac{y_{j+1} - y_j}{x_{j+1} - x_j} x + \frac{x_{j+1} y_j - x_j y_{j+1}}{x_{j+1} - x_j} - y_2)^2}, \qquad (4.8)$$

where (x_1, y_1) and (x_2, y_2) are the known locations of the radars. For ease in integration, substitute for the constants in equations (4.6)-(4.8) by

$$a = x_1, \qquad\qquad\qquad d = x_2,$$

$$b = \frac{y_{j+1} - y_j}{x_{j+1} - x_j}, \qquad\qquad e = \frac{x_{j+1} y_j - x_j y_{j+1}}{x_{j+1} - x_j} - y_2,$$

$$c = \frac{x_{j+1} y_j - x_j y_{j+1}}{x_{j+1} - x_j} - y_1, \qquad K = \sqrt{1 + \left(\frac{y_{j+1} - y_j}{x_{j+1} - x_j}\right)^2},$$

so the performance index is represented as

$$J_{1 \to 2} = K \int_{x_j}^{x_{j+1}} \left\{ \frac{\alpha_1}{[(x - a)^2 + (bx + c)^2]^2} + \frac{\alpha_2}{[(x - d)^2 + (bx + e)^2]^2} \right\} dx. \qquad (4.9)$$

This can be integrated through the use of tables or common symbolic mathematics software such as Mathematica to yield the following discrete approximation of the cost function,

$$J_{1 \to 2} = \frac{K \alpha_1}{2(ab + c)^3} \left[\frac{(ab + c)[(1 + b^2)x - a + bc]}{a^2 + c^2 - 2ax + 2bcx + (1 + b^2)x^2} - (1 + b^2) \operatorname{Arctan}(\frac{c + bx}{x - a}) \right]_{x_j}^{x_{j+1}} +$$

$$\frac{K \alpha_2}{2(db + e)^3} \left[\frac{(db + e)[(1 + b^2)x - d + be]}{d^2 + e^2 - 2dx + 2bex + (1 + b^2)x^2} - (1 + b^2) \operatorname{Arctan}(\frac{d + bx}{x - e}) \right]_{x_j}^{x_{j+1}}.$$

$$(4.10)$$

Thus the dependence upon x has been eliminated and the cost can be determined for any given pair of points (x_j, y_j) and (x_{j+1}, y_{j+1}). The total cost for a path of N line segments

4-5

is simply

$$\tilde{J}^* = \sum_{i=1}^{N} J_{i \to i+1} \, . \tag{4.11}$$

Again, this provides an accurate presentation of the continuous performance index, allowing the variational problem to be solved utilizing numerical techniques.

4.4 Application of the Euler Equation

The cost functional to be minimized is given in equation (4.5), repeated here for convenience

$$J = \int_0^{x_f} \left(\frac{\alpha_1}{R_1^4(x,y)} + \frac{\alpha_2}{R_2^4(x,y)} \right) \sqrt{1 + \dot{y}^2} \, dx \, . \tag{4.5}$$

Gelfand [10] identified this as a special case of problems where the desire is to minimize an integral of a function with respect to its arc length, i.e. a functional of the form

$$\int_a^b f(x,y) \sqrt{1 + \dot{y}^2} \, dx \, .$$

He proved that the Euler equation can be represented as

$$f_y - f_x \dot{y} - \frac{f}{1 + \dot{y}^2} \ddot{y} = 0 \, ,$$

or, solving for $\ddot{y}$,

$$\ddot{y} = \frac{1 + \dot{y}^2}{f} (f_x \dot{y} - f_y) \, , \tag{4.12}$$

where $f_x = \frac{\partial f}{\partial x}$ and $f_y = \frac{\partial f}{\partial y}$. For this problem, the function $f(x,y)$ is defined as

$$f(x,y) = \frac{\alpha_1}{R_1(x,y)^4} + \frac{\alpha_1}{R_2(x,y)^4} \, .$$

with derivatives

$$f_x = -\frac{4\,\alpha_1 \left(R_{1_x} + R_{1_y} \dot{y} \right)}{R_1{}^5} - \frac{4\,\alpha_2 \left(R_{2_x} + R_{2_y} \dot{y} \right)}{R_2{}^5} \, ,$$

$$f_y = -\frac{4\,R_{1_y}\,\alpha_1}{R_1{}^5} - \frac{4\,R_{2_y}\,\alpha_2}{R_2{}^5} \, .$$

4-6

Inserting f, f_x, and f_y into equation (4.12) and simplifying gives the Euler equation for the two radar case as

$$\ddot{y} = \frac{-4\left(1+\dot{y}\right)\alpha_1 R_2{}^5\left[\left(\dot{y}^2-1\right)R_{1_y}+\dot{y}\,R_{1_x}\right]-4\left(1+\dot{y}\right)\alpha_2 R_1{}^5\left[\left(\dot{y}^2-1\right)R_{2_y}+\dot{y}\,R_{2_x}\right]}{\alpha_2 R_1{}^5 R_2 + \alpha_1 R_1 R_2{}^5},$$

(4.13)

where equations (4.3) and (4.4) provide us R_1, R_2 and their partial derivatives,

$$R_1 = \sqrt{(x-x_1)^2+(y-y_1)^2}\,, \qquad\qquad R_2 = \sqrt{(x-x_2)^2+(y-y_2)^2}\,,$$

$$R_{1_x} = \frac{x-x_1+(y-y_1)\,\dot{y}}{\sqrt{(x-x_1)^2+(y-y_1)^2}}\,, \qquad R_{2_x} = \frac{x-x_2+(y-y_2)\,\dot{y}}{\sqrt{(x-x_2)^2+(y-y_2)^2}}\,,$$

$$R_{1_y} = \frac{y-y_1}{\sqrt{(x-x_1)^2+(y-y_1)^2}}\,, \qquad R_{2_y} = \frac{y-y_2}{\sqrt{(x-x_2)^2+(y-y_2)^2}}\,.$$

4.5 Trajectory Optimization Against Two Equal Power Radars

This section examines the case of air vehicle flight against two equal power radars. A short development of the Voronoi path is presented first, followed by the results of the optimization.

4.5.1 Voronoi Comparison Path for $\alpha_1 = \alpha_2$. For this study, vehicle flight was against at most two radars; thus, given two equal power radars located at (x_1, y_1) and (x_2, y_2), the perpendicular bisector of the line segment connecting the radars will be the Voronoi edge.

The equation of the line connecting the radars is

$$y(x) = m(x - x_2) + y_2\,,$$

(4.14)

$$m = \frac{y_2 - y_1}{x_2 - x_1}\,.$$

(4.15)

The perpendicular bisector of a line has slope $-1/m$ and passes through the midpoint of the line (x_m, y_m),

$$y_\perp(x) = -\frac{1}{m}(x - x_m) + y_m \,, \tag{4.16}$$

$$x_m = \frac{x_1 + x_2}{2} \,, \tag{4.17}$$

$$y_m = \frac{y_1 + y_2}{2} \,. \tag{4.18}$$

The comparison path will be constructed of three line segments: a shortest path line from the initial point, the perpendicular bisector, and the shortest path line to the final point completing the curve, shown in Figure 4.3. The first intercept (x_{i1}, y_{i1}) is on the line perpendicular to (4.16) through (x_0, y_0),

$$\tilde{y}(x) = m(x - x_0) + y_0 \,. \tag{4.19}$$

Equating $\tilde{y}(x)$ to (4.16) and solving for x yields

$$x_{i1} = \frac{m^2 x_0 + m(y_m - y_0) + x_m}{m^2 + 1} \,, \tag{4.20}$$

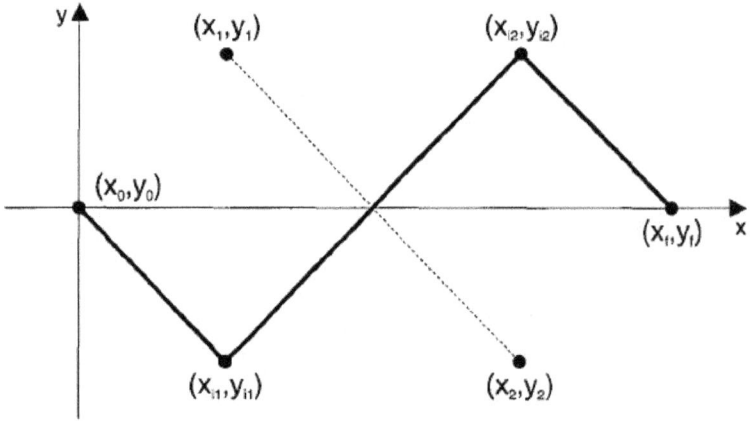

Figure 4.3   Voronoi Path for Radars of Equal Transmission Power, $\alpha_1 = \alpha_2$

and back-substituting into $\tilde{y}$ gives

$$y_{i1} = \frac{m^2 y_m + m(x_m - x_0) + y_0}{m^2 + 1} . \tag{4.21}$$

Taking a similar approach at the final point of the path, (x_f, y_f), results in

$$x_{i2} = \frac{m^2 x_f + m(y_m - y_f) + x_m}{m^2 + 1} , \tag{4.22}$$

$$y_{i2} = \frac{m^2 y_m + m(x_m - x_f) + y_f}{m^2 + 1} , \tag{4.23}$$

where m, x_m, and y_m are given by (4.15), (4.17), and (4.18).

4.5.2   Optimization Results.   For individual plots of the trajectories generated, see Appendices A.1-A.3.

4.5.2.1   Scenario 1: Varying Downrange Radar Separation.   The first case explores changing the downrange location of the radars and the effect on the optimal trajectory. Six optimal trajectories were produced corresponding to six different symmetric radar geometries. A summary of the optimal trajectory cost, J^*, and path length, ℓ^*, the Voronoi path cost, J_{vor}, and path length, ℓ_{vor}, and the straight line cost, J_{line}, and path length, ℓ_{line} is presented in Table 4.2.

Intuitively, it is expected that the paths will be symmetric about the midpoint of the vector connecting the radars, and the path will bend away from the nearest radar. The results of the optimization do in fact prove this to be true. In Figure 4.4, the optimal path

Table 4.2   Objective Cost, J, and Path Length, ℓ, for Scenario 1, $\alpha_1 = \alpha_2$

A	J^*	ℓ^*	J_{vor}	ℓ_{vor}	J_{line}	ℓ_{line}
0.0	20.566371	1.000000	20.566371	1.000000	20.566371	1.000000
0.2	19.644590	1.018444	20.698733	1.176697	20.243257	1.000000
0.4	17.363391	1.065750	18.677842	1.299867	19.240926	1.000000
0.6	14.656035	1.120135	15.933684	1.371989	17.495753	1.000000
0.8	12.108780	1.155987	13.296891	1.405564	15.023435	1.000000
1.0	9.797058	1.159667	10.851193	1.414214	12.057190	1.000000

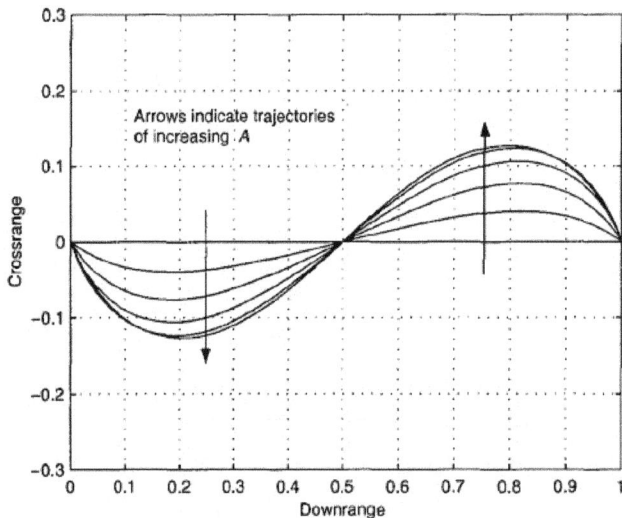

Figure 4.4   Optimal Trajectories for Increasing A, $\alpha_1 = \alpha_2$

bends away from the nearest radar and intersects the downrange axis at the midpoint of the radars as the radars move towards the endpoints. From Table 4.2, the path length of the optimal path is increasing as A increases. Interestingly, at the point the radars are at the same x-coordinate as the endpoints, the path length is shorter than the previous trajectory. This is because as A becomes greater than C, the optimal trajectory will approach a straight line which is short compared to magnitude of A. As $A \to \infty$, the ratio of optimal path length to downrange distance between radars, C/A, and the optimal cost, J^*, will approach zero. This trend is evident in Figure 4.5, with the cost approaching zero as the downrange distance between radars increases. Therefore, the maximum value for the objective cost of the optimal trajectory is when $C/A \to \infty$, or when $A \to 0$.

The relation of the optimal path to the Voronoi path is not evident from this scenario. It appears at first that the optimal curve is a smoothed function of the Voronoi path; or, viewed from another perspective, the Voronoi path is a rough linear approximation to the optimal curve. In the limit as A gets very large, however, the Voronoi edge will become perpendicular to the optimal path and the Voronoi intercepts will move to a single point, the midpoint of the Voronoi edge between the radars. The only consistently common point is the midpoint of the paths, and this is due to the symmetry of the problem. Thus, graphically there appears to be little similarity between the Voronoi path and the optimal

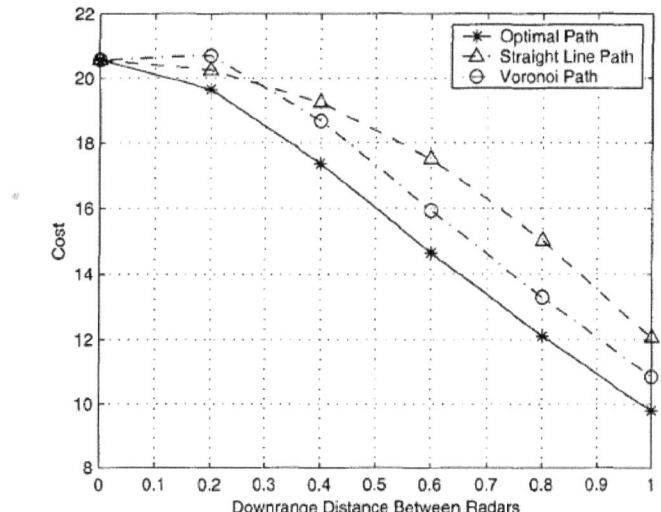

Figure 4.5   Objective Cost as a Function of Downrange Distance, $\alpha_1 = \alpha_2$

path. Comparing the objective cost of the three paths, however, the cost of the Voronoi path is mainly between the optimal cost and the straight line cost. This indicates that while it is a suboptimal path, it is a better approximation of the optimal path than the straight line.

4.5.2.2   Scenario 2: Varying Crossrange Radar Separation.   In Scenario 2, the radars are kept at a fixed downrange separation, A, while the crossrange, B, is progressively increased, Table 4.3. Results similar to what was observed in Scenario 1 are expected;

Table 4.3   Objective Cost, J, and Path Length, ℓ, for Scenario 2, $\alpha_1 = \alpha_2$

B	J^*	ℓ^*	J_{vor}	ℓ_{vor}	J_{line}	ℓ_{line}
0.4	–	–	139.600038	1.299867	76.497903	1.000000
0.6	35.415026	1.413575	41.930698	1.371989	57.576626	1.000000
0.8	17.383209	1.254360	19.308828	1.405564	23.986211	1.000000
1.0	9.797921	1.159563	10.851193	1.414214	12.057190	1.000000
1.2	5.964369	1.102037	6.727189	1.408406	6.812747	1.000000
1.4	3.826028	1.066506	4.409475	1.394972	4.168714	1.000000
1.6	2.555771	1.044101	3.000367	1.377997	2.702852	1.000000
1.8	1.765128	1.029661	2.100850	1.359800	1.831576	1.000000
2.0	1.253887	1.020306	1.506437	1.341641	1.285398	1.000000

as the crossrange $B \to \infty$, the optimal path will approach a straight line. Indeed, this is observed in the optimal trajectories for this case, Figure 4.6.

For this scenario, numerical difficulties preempted finding solutions as $B \to 0$. This is likely due to the optimal path desiring to travel around the radars instead of between them. When problems occurred, the calculated solution would travel through one of the radars. This is because the solution of the Euler equation provides an extremum, not just a minimum; these cases were obviously maximum or saddle point solutions to the Euler equation.

From Figure 4.7 inferences can still be drawn as to the affects of varying B. For this formulation, as $B \to 0$, the cost, J^*, will approach some very large number. Since x must monotonically increase, the path has nowhere to go but through the radars. In reality, the optimal trajectory would never follow this path; instead, it would travel around the radars at a much lower cost. As mentioned before, to solve this problem an alternate formulation is required.

When B is increased, J^* will go to zero and the optimal path will become a straight line. In addition, the Voronoi path will also flatten to a straight line. From Figure 4.7, a

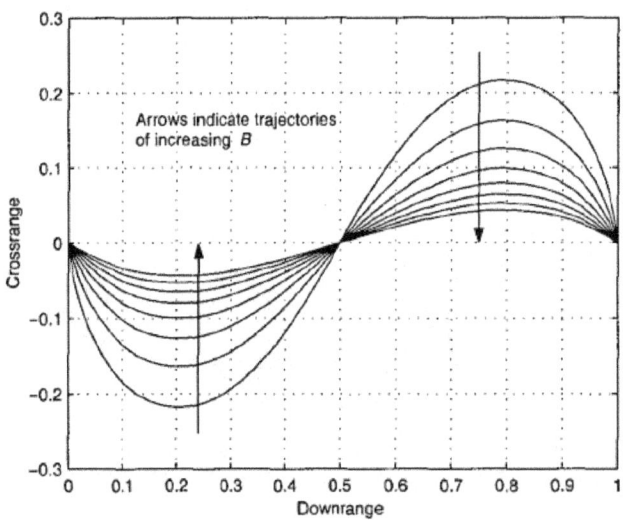

Figure 4.6   Optimal Trajectories for Increasing B, $\alpha_1 = \alpha_2$

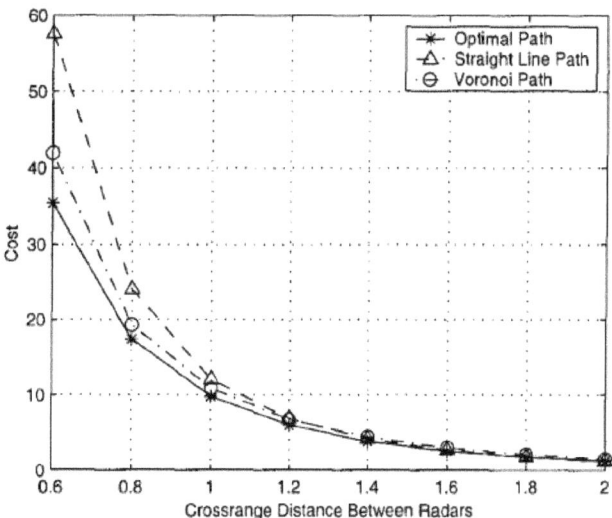

Figure 4.7   Objective Cost as a Function of Crossrange Distance, $\alpha_1 = \alpha_2$

straight line path is nearly as effective as the optimal path and better than the Voronoi path when $B/C \approx 1.6/1$, and at that ratio there is a 4% savings in path length.

The results of the first two scenarios followed the expectations of how the optimal trajectory would react to different radar geometries, and reinforced the fact that the path will bend away from the radars when they are close and approach the direct path as the radars move away from the endpoints. Little information could be gleaned by comparing the optimal trajectory to the Voronoi path, since it changed with each iteration. In Scenario 3, the radar geometry is held fixed and the path endpoint separation is progressively increased. Since the radars do not move, the Voronoi edge will be constant and the only variable in the Voronoi path will be the length of the segments connecting the endpoints. The resulting optimal trajectories reveal an opportunity to exploit the Voronoi path for on-line utilization.

4.5.2.3   Scenario 3:   Varying Downrange Endpoint Separation.   For this scenario, the distance between the path's endpoints are iteratively increased while the radar locations are fixed. This means that the Voronoi edge is constant, and the Voronoi path just gets progressively longer as the endpoint separation gets large. Table 4.4 and Figure 4.8 summarize the optimal trajectories calculated.

Table 4.4   Objective Cost, J, and Path Length, ℓ, for Scenario 3, $\alpha_1 = \alpha_2$

C	J^*	ℓ^*	J_{vor}	ℓ_{vor}	J_{line}	ℓ_{line}
0.2	5.800936	0.200057	6.747879	0.235339	5.802672	0.200000
0.4	10.932462	0.401157	12.380733	0.470679	10.970394	0.400000
0.6	14.937594	0.604842	16.429929	0.706018	15.103038	0.600000
0.8	17.763680	0.810947	19.065072	0.941357	18.135953	0.800000
1.0	19.645656	1.018353	20.698733	1.176697	20.243257	1.000000
1.2	20.874626	1.226197	21.701820	1.412036	21.671344	1.200000
1.4	21.681834	1.433928	22.326437	1.647376	22.635972	1.400000
1.6	22.221526	1.641400	22.725487	1.882715	23.293968	1.600000
1.8	22.590806	1.848554	22.988123	2.118054	23.750271	1.800000
2.0	22.849767	2.055328	23.166208	2.353394	24.072858	2.000000
3.0	23.410955	3.085858	23.529910	3.530090	24.776987	3.000000
4.0	23.569667	4.113160	23.625032	4.706787	24.975693	4.000000
5.0	23.629938	5.139262	23.659745	5.883484	25.050604	5.000000
6.0	23.657501	6.164908	23.675263	7.060181	25.084653	6.000000
7.0	23.671810	7.190211	23.683204	8.236878	25.102245	7.000000
8.0	23.679951	8.215458	23.687679	9.413574	25.112217	8.000000
9.0	23.684914	9.240641	23.690389	10.590271	25.118281	9.000000
10.0	23.688106	10.265973	23.692125	11.766968	25.122176	10.000000

The solutions for the optimal trajectory seem alarming at first. As the endpoints move outward from the radars, the path extends further and further out. While the

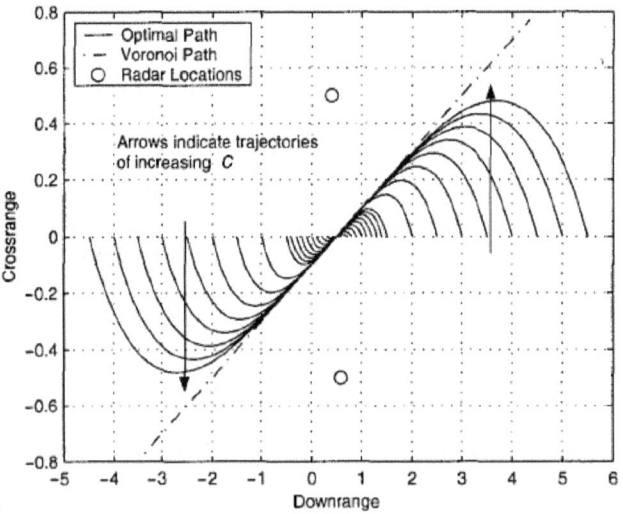

Figure 4.8   Optimal Trajectories for Increasing C, $\alpha_1 = \alpha_2$

4-14

optimal trajectories seem excessive, looking at the relationship between objective cost and the endpoint separation C, Figure 4.9, it is seen that the cost for a straight line and the optimal path are relatively close, and a straight line path would be acceptable. Upon further review, a more important result is uncovered: the cost of the Voronoi path is nearly identical to the optimal path. In fact, after $C > 1.4$, the difference in cost is essentially negligible. Thus, the question is no longer how to get from the initial point to the final point; the question now is how to optimally approach to the Voronoi edge from the initial point and how to optimally depart the Voronoi edge to get to the final point. This has crucial implications for on-line path planning. Instead of a full path optimization being performed, utilizing valuable on-line system resources, one only needs to optimize the approach and departure from the Voronoi edge. Research in this area is currently being performed at the Air Force Research Laboratory.

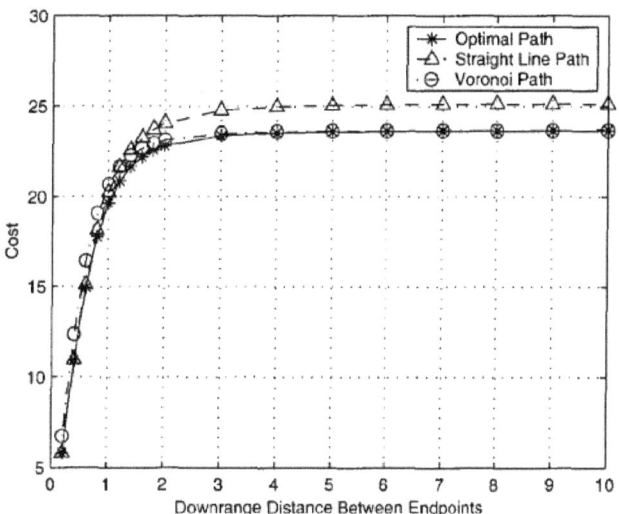

Figure 4.9   Objective Cost as a Function of Endpoint Separation, $\alpha_1 = \alpha_2$

4.6   Trajectory Optimization Against Two Unequal Power Radars

4.6.1   Voronoi Comparison Path for $\alpha_1 \neq \alpha_2$.
For the case when the radars are of unequal transmission power, a weighted Voronoi diagram is used. The resulting curve is known as a circle of Apollonius. Okabe [25] provides a full derivation of the theory; a simple Cartesian formulation for this specific study is provided here. The weighted Voronoi

path starts with equating the power received by each radar, (4.1),

$$\frac{\alpha_1}{R_1^4} = \frac{\alpha_2}{R_2^4} \tag{4.24}$$

$$\Rightarrow \alpha_1 R_2^4 = \alpha_2 R_1^4 \tag{4.25}$$

$$\Rightarrow \pm\sqrt{\alpha_1} R_2^2 = \pm\sqrt{\alpha_1} R_1^2, \tag{4.26}$$

where R_1 and R_2 are defined in (4.3)-(4.4). From (4.26) it appears four cases must be investigated; however, two of the cases are equivalent to each other and can be eliminated. The two cases to examine are

$$\sqrt{\alpha_1} R_2^2 = \sqrt{\alpha_1} R_1^2, \tag{4.27}$$

$$\sqrt{\alpha_1} R_2^2 = -\sqrt{\alpha_1} R_1^2, \tag{4.28}$$

$$\alpha_1 \geq 1, \ \alpha_2 \geq 1, \ \alpha_1 \neq \alpha_2. \tag{4.29}$$

We will begin by molding equation (4.27) into a more familiar form. To simplify the notation, let $\sqrt{\alpha_1} = a$, $\sqrt{\alpha_2} = b$. Substituting R_1 and R_2 from equations (4.3)-(4.4) into equation (4.27) gives

$$a[(x - x_2)^2 + (y - y_2)^2] = b[(x - x_1)^2 + (y - y_1)^2]. \tag{4.30}$$

Now let c = b/a, and expand and collect terms,

$$x^2(1 - c) - 2x(x_2 - cx_1) + y^2(1 - c) - 2y(y_2 - cy_1) = (cx_1^2 - x_2^2) + (cy_1^2 - y_2^2). \tag{4.31}$$

Divide through by $(1 - c)$,

$$x^2 - 2x\frac{x_2 - cx_1}{1 - c} + y^2 - 2y\frac{y_2 - cy_1}{1 - c} = \frac{cx_1^2 - x_2^2 + cy_1^2 - y_2^2}{1 - c}. \tag{4.32}$$

Noticing that we can complete the square for the x- and y-terms gives

$$\left(x - \frac{x_2 - cx_1}{1 - c}\right)^2 - \left(\frac{x_2 - cx_1}{1 - c}\right)^2 + \left(y - \frac{y_2 - cy_1}{1 - c}\right)^2 - \left(\frac{y_2 - cy_1}{1 - c}\right)^2 = \frac{cx_1^2 - x_2^2 + cy_1^2 - y_2^2}{1 - c}. \tag{4.33}$$

Moving the constant terms to the right-hand side and simplifying yields

$$\left(x - \frac{x_2 - cx_1}{1 - c}\right)^2 + \left(y - \frac{y_2 - cy_1}{1 - c}\right)^2 = \left(\frac{\sqrt{c}}{1 - c}\sqrt{(x_2 - x_1)^2 + (y_2 - y_2)^2}\right)^2, \tag{4.34}$$

which is obviously the equation of a circle. Applying an identical formulation for the second case, equation (4.28), gives the following equation of a circle

$$\left(x - \frac{x_2 + cx_1}{1 + c}\right)^2 + \left(y - \frac{y_2 + cy_1}{1 + c}\right)^2 = \left(\frac{i\sqrt{c}}{1 + c}\sqrt{(x_2 - x_1)^2 + (y_2 - y_2)^2}\right)^2. \tag{4.35}$$

Since $c = \sqrt{\alpha_2/\alpha_1} > 0$, equation (4.35) has a radius which exists in the imaginary realm; hence, equation (4.35) will be discarded and equation (4.34) will be used. Substituting for c, the locus of equal power received for radars of unequal transmission power can be summarized as a circle with center (x_c, y_c) and radius r_c,

$$x_c = \frac{\sqrt{\alpha_1}x_2 - \sqrt{\alpha_2}x_1}{\sqrt{\alpha_1} - \sqrt{\alpha_2}}, \tag{4.36}$$

$$y_c = \frac{\sqrt{\alpha_1}y_2 - \sqrt{\alpha_2}y_1}{\sqrt{\alpha_1} - \sqrt{\alpha_2}}, \tag{4.37}$$

$$r_c = \frac{\sqrt{\sqrt{\alpha_1 \alpha_2}}}{\sqrt{\alpha_1} - \sqrt{\alpha_2}}\sqrt{(x_2 - x_1)^2 + (y_2 - y_2)^2}. \tag{4.38}$$

A comparison path similar to the perpendicular bisector will be constructed using equations (4.36)-(4.38), with the shortest path taken from the endpoints to the Voronoi edge completing the curve as before, Figure 4.10.

The first intercept (x_{i1}, y_{i1}) is on the line perpendicular to (4.34) through (x_0, y_0) and (x_c, y_c),

$$\tilde{y} = \frac{y_c - y_0}{x_c - x_0}(x - x_0) + y_0. \tag{4.39}$$

4-17

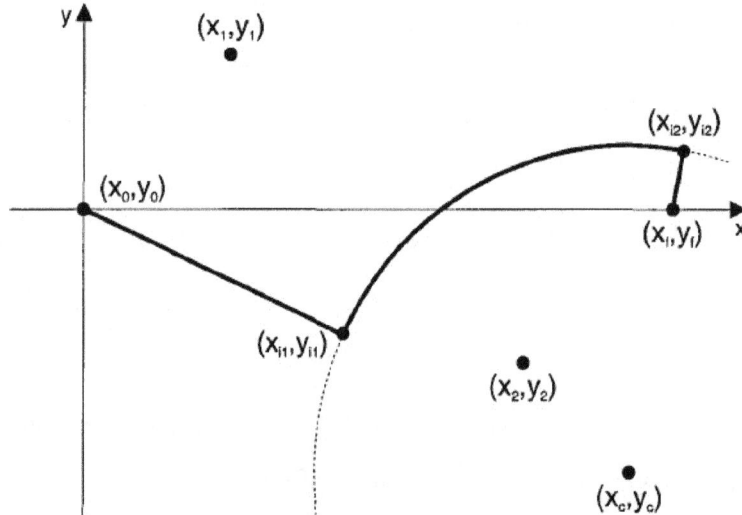

Figure 4.10   Voronoi Path for Radars of Unequal Transmission Power, $\alpha_1 \neq \alpha_2$

Substituting $y = \tilde{y}$ in (4.34)

$$(x - x_c)^2 + \left(\frac{y_c - y_0}{x_c - x_0}(x - x_0) + y_0 - y_c\right)^2 = r_c^2, \qquad (4.40)$$

and solving for x gives

$$x_{i1} = x_c \pm \frac{r_c(x_c - x_0)}{\sqrt{(x_c - x_0)^2 + (y_c - y_0)^2}}. \qquad (4.41)$$

To determine y_{i1}, substitute x_{i1} into equation (4.39) and simplify,

$$y_{i1} = y_c \pm \frac{r_c(y_c - y_0)}{\sqrt{(x_c - x_0)^2 + (y_c - y_0)^2}}. \qquad (4.42)$$

This gives the four intercept points of the lines through the endpoints with the circle. For this study, $\alpha_1 > \alpha_2$ and $\alpha_2 = 1$; with the geometry of Figure 4.10, this placed (x_c, y_c) in the fourth quadrant. Since we are interested in only the points between the radars, the intercept point is at

$$x_{i1} = x_c - \frac{r_c(x_c - x_0)}{\sqrt{(x_c - x_0)^2 + (y_c - y_0)^2}},$$ (4.43)

$$y_{i1} = y_c - \frac{r_c(y_c - y_0)}{\sqrt{(x_c - x_0)^2 + (y_c - y_0)^2}}.$$ (4.44)

4.6.2 Optimization Results.   Optimal trajectories for the three scenarios are again calculated, but now with unequal radar transmission power. The ratio of transmission powers is $\alpha_1/\alpha_2 = 2/1$, where α_1 is the power of the radar at (x_1, y_1) and α_2 is the power of the radar at (x_2, y_2). For comparison purposes, the ordinary Voronoi path is replaced by its equivalent weighted Voronoi path for the radar geometry. For individual plots of the trajectories generated, see Appendices A.4-A.6.

4.6.2.1 Scenario 1: Varying Downrange Radar Separation.   Table 4.5 is a summary of the optimal cost and path lengths for this scenario. Representative trajectories are shown in Figures 4.11-4.12, and all of the optimal paths in Figure 4.13. As expected, the optimal trajectories behaved similarly to the case when the radars were of equal power. The optimal path is now asymmetric, and bends further away from the radar in the first quadrant because it is radiating with twice the power of the other radar. The path no longer intercepts the midpoint of the vector connecting the radars; instead the intercept point lies where that vector intercepts the locus of equal power. The optimal trajectory bends just enough to closely follow the weighted Voronoi path for a short while and then leaves the path to meet the endpoint constraint.

Table 4.5   Objective Cost, J, and Path Length, ℓ, for Scenario 1, $\alpha_1 \neq \alpha_2$

A	J^*	ℓ^*	J_{vor}	ℓ_{vor}	J_{line}	ℓ_{line}
0.0	29.640530	1.008353	31.221704	1.147263	30.849556	1.000000
0.2	28.331782	1.026904	29.594596	1.151912	30.364886	1.000000
0.4	25.098595	1.074275	26.925237	1.278581	28.861389	1.000000
0.6	21.271201	1.128260	23.180429	1.355481	26.243630	1.000000
0.8	17.671044	1.163131	19.506644	1.393091	22.535153	1.000000
1.0	14.381110	1.165595	16.011071	1.404509	18.085785	1.000000

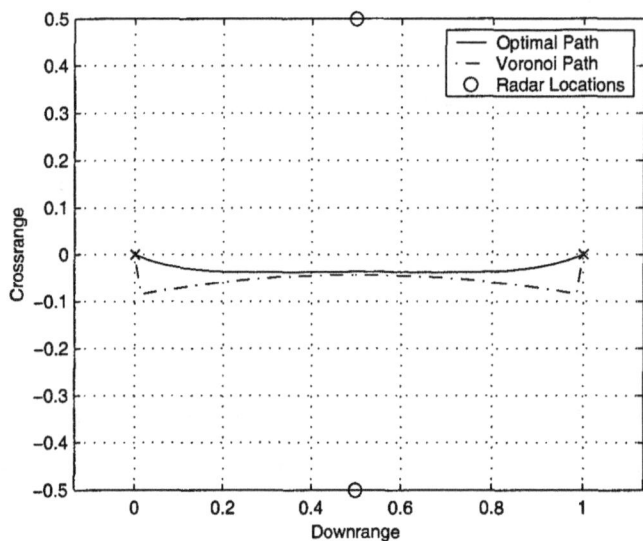

Figure 4.11   Optimal Trajectory for $\alpha_1/\alpha_2 = 2/1$

The shape of the curve describing the objective cost as a function of A, Figure 4.14, is nearly the same as in Section 4.5.2.1. The magnitude of the curve for this case is higher, but this is expected as the vehicle is constrained to travel in the same area as before but with one higher-powered radar. The relationship between A and C remains the same: as

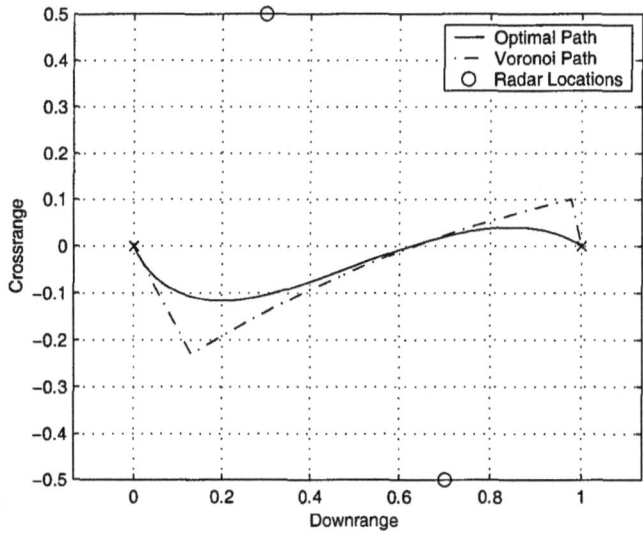

Figure 4.12   Optimal Trajectory for $\alpha_1/\alpha_2 = 2/1$

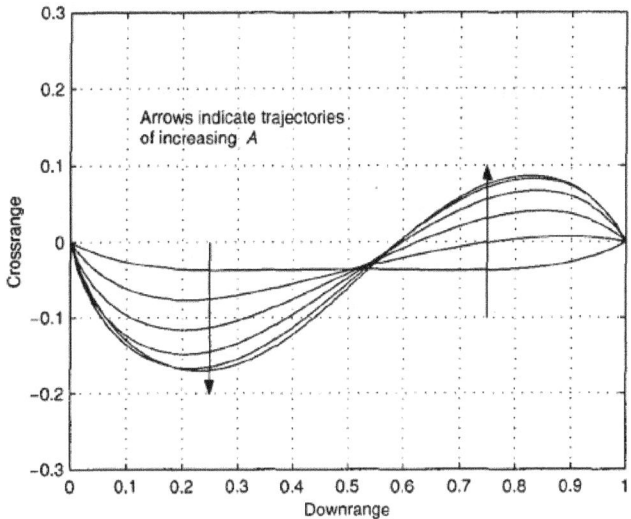

Figure 4.13   Optimal Trajectories for Increasing A, $\alpha_1 \neq \alpha_2$

C/A approaches infinity, J^* will be at its maximum, and as C/A approaches zero, J^* will approach zero. As was the case in Section 4.5.2.1, one cannot easily discern the relationship between the optimal path and the Voronoi path from this scenario because the Voronoi path is changing with each iteration.

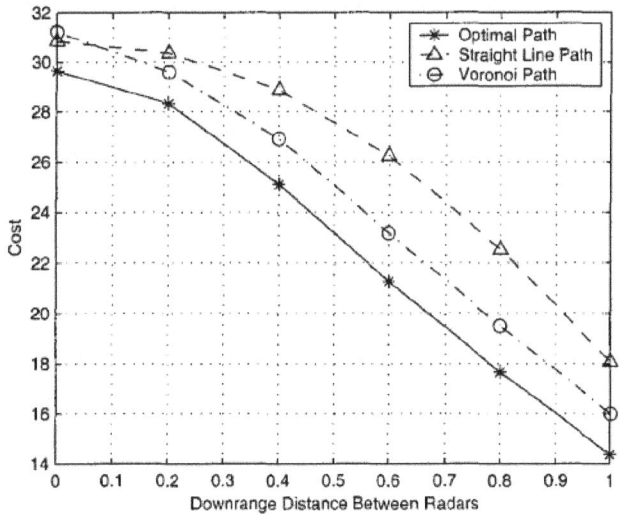

Figure 4.14   Objective Cost as a Function of Radar Downrange Separation, $\alpha_1 \neq \alpha_2$

4.6.2.2 Scenario 2: Varying Crossrange Radar Separation. The trajectories generated for the case when the radars are of unequal power track the same trends explained in Section 4.5.2.2, as shown in Table 4.6 and Figures 4.15-4.16. The optimal path is again asymmetric, as in Section 4.6.2.1, bending away from the radar of greater power. The path length of the optimal trajectories are only slightly longer than those of Section 4.5.2.2, at most 1% greater. To make an analogy, if the optimal paths from Section 4.5.2.2 were strings of constant length, the paths of this section are merely a repositioning of that string. This is likely a factor of the radar power ratio and the length difference may become more pronounced as the ratio increases.

As B is increased, the objective cost follows nearly the same curve as the equal radar power scenario, only translated upward in cost, due to the higher-power radar. Similar to the results in Section 4.5.2.2, when the ratio $B/C = 1.6/1$, there is only a 1% difference in cost between the straight line trajectory and the optimal path, while the straight line path is 5% shorter than the optimal. Depending upon the application, the shorter travel distance may outweigh the minor increase in radar exposure.

Numerical difficulties again prevented the calculation of optimal trajectories as B approached the abscissa. As was the case before, the optimal trajectory would likely go around the radars and could be solved for using an alternate formulation.

Table 4.6   Objective Cost, J, and Path Length, ℓ, for Scenario 2, $\alpha_1 \neq \alpha_2$

B	J^*	ℓ^*	J_{vor}	ℓ_{vor}	J_{line}	ℓ_{line}
0.4	–	–	211.271635	1.300402	152.015541	1.000000
0.6	–	–	63.063121	1.369012	47.172170	1.000000
0.8	25.577900	1.259919	28.749046	1.398834	35.979317	1.000000
1.0	14.381110	1.165595	16.011071	1.404509	18.085785	1.000000
1.2	8.744103	1.108858	9.858671	1.396655	10.219120	1.000000
1.4	5.607863	1.074068	6.430085	1.381909	6.253072	1.000000
1.6	3.747165	1.052340	4.359243	1.364122	4.054278	1.000000
1.8	2.589466	1.038537	3.043824	1.345432	2.747364	1.000000
2.0	1.841082	1.029601	2.177832	1.326976	1.928097	1.000000

4.6.2.3 Scenario 3: Varying Downrange Endpoint Separtion. As in Section 4.5.2.3, by varying the parameter C the best approach for travelling between the radars

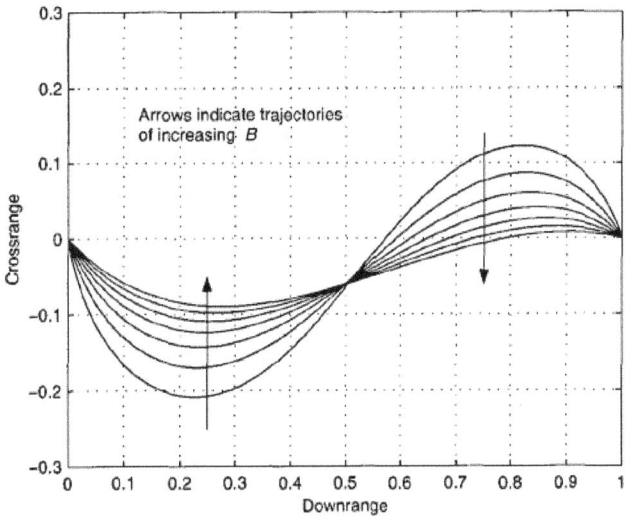

Figure 4.15   Optimal Trajectories for Increasing B, $\alpha_1 \neq \alpha_2$

can be determined. The results of the optimization are shown in Table 4.7 and Figures 4.17-4.18. For equal radar power, it was shown that the optimal path can be determined by optimizing the approach to and departure from the Voronoi edge. When the radars are of unequal transmission power, a similar phenomenon takes place. Whereas in Section 4.5.2.3 the optimal trajectory approached the perpendicular bisector, for this scenario the

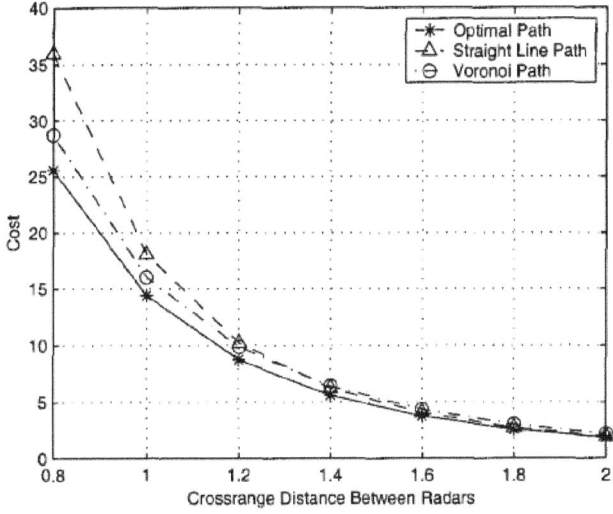

Figure 4.16   Objective Cost as a Function of Radar Crossrange Separation, $\alpha_1 \neq \alpha_2$

4-23

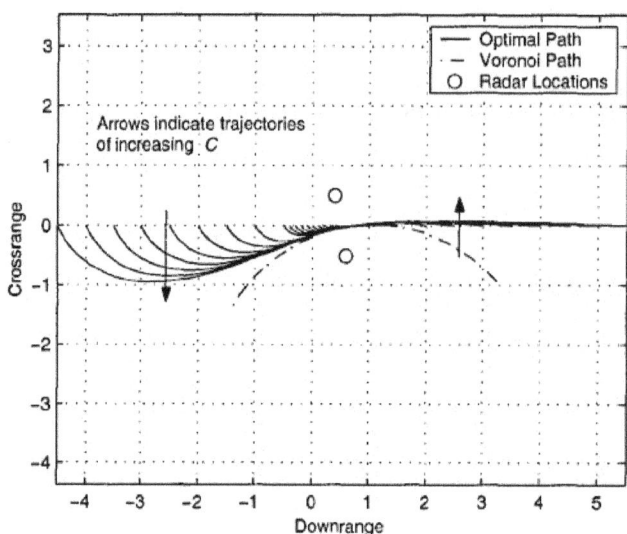

Figure 4.17   Optimal Trajectories for Increasing C, $\alpha_1 \neq \alpha_2$

optimal trajectory approached the weighted Voronoi edge, a circular locus of equal radar power. At the intersection with the vector connecting the radars, the optimal trajectory departed the locus and turned towards the final point. Interestingly, the final portion of the path does not follow the same trend as in the equal power case. The optimal trajectory breaks off of the Voronoi edge much sooner, indicating that it is less costly to get to the final point quickly than to follow the locus of equal power. As concluded in Section 4.5.2.3, the problem to be solved now is how to optimally approach and depart from the Voronoi edge. For on-board usage, this has a much lower computational cost than performing a full trajectory optimization problem.

Table 4.7   Objective Cost, J, and Path Length, ℓ, for Scenario 3, $\alpha_1 \neq \alpha_2$

C	J^*	ℓ^*	J_{vor}	ℓ_{vor}	J_{line}	ℓ_{line}
0.2	8.625837	0.201447	9.640721	0.284520	8.704008	0.200000
0.4	16.044973	0.405500	17.692042	0.486992	16.455591	0.400000
0.6	21.733376	0.610855	23.483481	0.695163	22.654557	0.600000
0.8	25.707102	0.818134	27.254841	0.924455	27.203930	0.800000
1.0	28.332287	1.026878	29.594596	1.151912	30.364886	1.000000
1.2	30.034460	1.236385	30.825946	1.376988	32.507016	1.200000
1.4	31.144935	1.446190	31.813057	1.599261	33.953959	1.400000
1.6	31.883008	1.656081	32.465301	1.818338	34.940951	1.600000
1.8	32.385425	1.865962	32.903663	2.040986	35.625407	1.800000
2.0	32.736111	2.075784	33.203846	2.277616	36.109287	2.000000
3.0	33.488869	3.125072	33.807335	3.490995	37.165481	3.000000
4.0	33.698376	4.175579	33.948015	4.727328	37.463540	4.000000
5.0	33.777111	5.227868	33.991086	5.960318	37.575907	5.000000
6.0	33.813018	6.281839	34.006707	7.175999	37.626979	6.000000
7.0	33.831137	7.337236	34.012977	8.368961	37.653367	7.000000
8.0	33.841803	8.393788	34.015608	9.538646	37.668326	8.000000
9.0	33.848026	9.451291	34.016680	10.686775	37.677422	9.000000
10.0	33.852026	10.509581	34.017028	11.815895	37.683265	10.000000

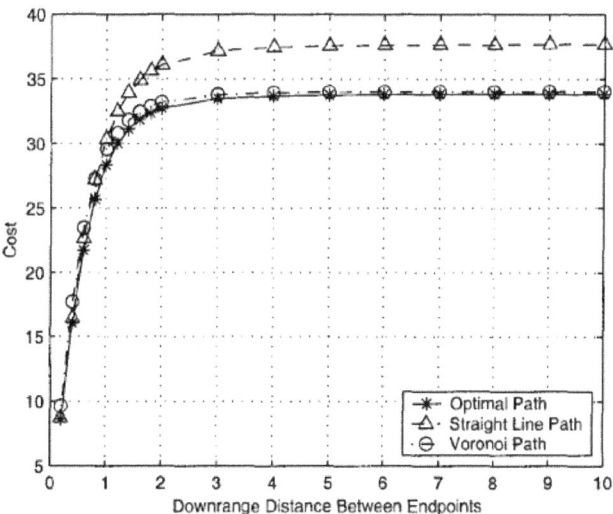

Figure 4.18   Objective Cost as a Function of Endpoint Separation, $\alpha_1 \neq \alpha_2$

V. Conclusions and Recommendations

5.1 Conclusions

The application of different numerical methods to optimizing the trajectory of an air vehicle against one radar highlighted the necessity of understanding the fundamental problem. The capability for comparison with the analytical baseline was invaluable in understanding the dynamics of the optimal trajectories as they approached the theoretical limit. For several of the cases, the numerical optimization found trajectories that were excessively long yet having a cost very near the optimal. Without a priori knowledge of the angle limitations, these trajectories could have mistakenly been classified as optimal. Instead, it led to exploration of the relationship between the objective cost and path length, and the finding that radar exposure minimization against one radar is relatively invariant to path length. This is an advantageous result, as it means that sub-optimal paths of shorter path length can be substituted for the optimal path at very little risk. This could lead to reduced flight times and fuel savings, two of the more important parameters for operational mission planners.

Through the systematic variation of the downrange and crossrange radar separation, the shaping of the optimal trajectory with respect to varying radar geometries was identified. Additionally, varying the crossrange radar separation helped identify situations where different formulations or numerical methods might be necessary. By exploring the effect of these parameters, a more robust numerical optimization technique can be developed.

In contrast, varying the path endpoint separation provided a way to analyze travelling through the radars. Through direct comparison of objective cost and path length with the Voronoi diagram, it was discovered that for a feasible near-optimal trajectory, one option is to optimize the approach to and departure from the Voronoi edge. The calculation of the ordinary or weighted Voronoi edge is a simple task, and if through additional research a similarly deterministic calculation of the approach and departure paths is found, the resulting suboptimal trajectory can be quickly calculated such that it approximates the optimal solution to within acceptable limits. This is a significant result, especially for on-line optimization. Partial path optimization is a much quicker and cheaper on-line calculation than performing a full path optimization, thus conserving valuable system resources for other mission tasks.

5.2 Recommendations For Future Research

For a full understanding of the air vehicle flight optimization problem, it is necessary to examine what happens if the air vehicle can only fly around the radars and not between them. A reformulation of the problem in which no constraint exists on the solution space would eliminate the numerical difficulties encountered in this study and open the problem for additional examination. Calculation of optimal trajectories with this constraint would provide important information as to when it is more advantageous to travel between versus around the radars.

Removing the travel path restriction and adding a path length constraint would allow for a Pareto analysis of the objective cost versus the path length, similar to the analysis done in Chapter 3 of this document. This would help determine the proverbial "knee-in-the-curve" at which additional path length gives little risk savings, an important consideration when time and fuel usage are of concern.

An analogous assessment of air vehicle trajectory optimization might uncover ways to effectively reduce the amount of received radar power by bistatic radar systems. Bistatic radar systems utilize transmitters and receivers at different locations, providing the ability to track smaller radar cross section aircraft. The extension of this problem to encompass bistatic radars would be an informative study and may provide additional insights for both radar types.

Finally, a generalization of the problem to any number of radars is necessary for true on-board capability to be realized. In a radar-rich environment, the air vehicle avoiding detection needs to have a quick computational method for discerning the best possible trajectory to complete the mission. Application of the results of this and studies encompassing the recommendations above could be fused into a deterministic calculation of the optimal path, conserving system resources for other mission-essential tasks.

Appendix A. Plots for Trajectory Optimization Against Two Radars

A.1 Scenario 1: Varying Downrange Radar Separation, $\alpha_1 = \alpha 2$

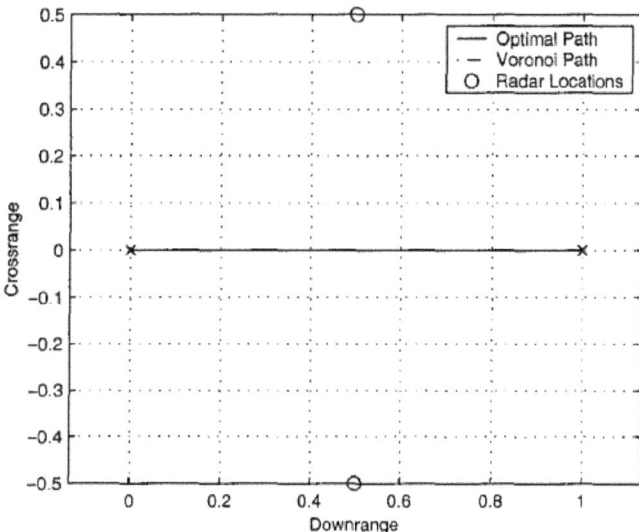

Figure A.1   Optimal Trajectory for $A = 0.0$

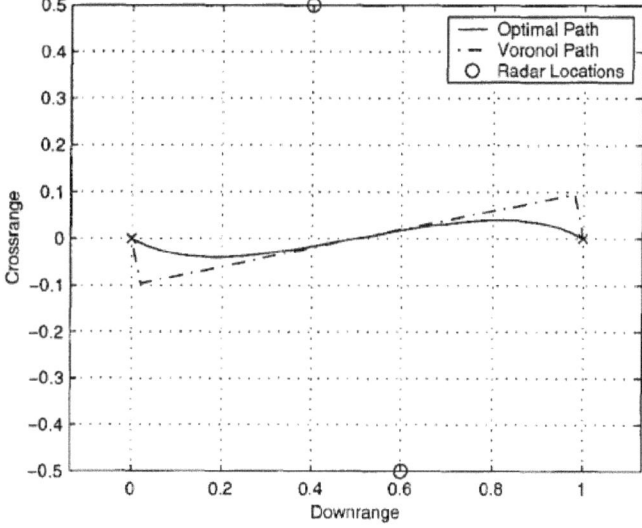

Figure A.2   Optimal Trajectory for $A = 0.2$

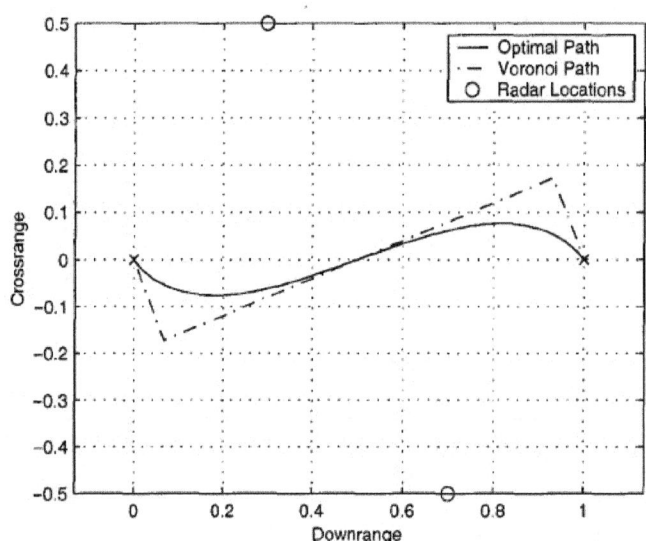

Figure A.3   Optimal Trajectory for $A = 0.4$

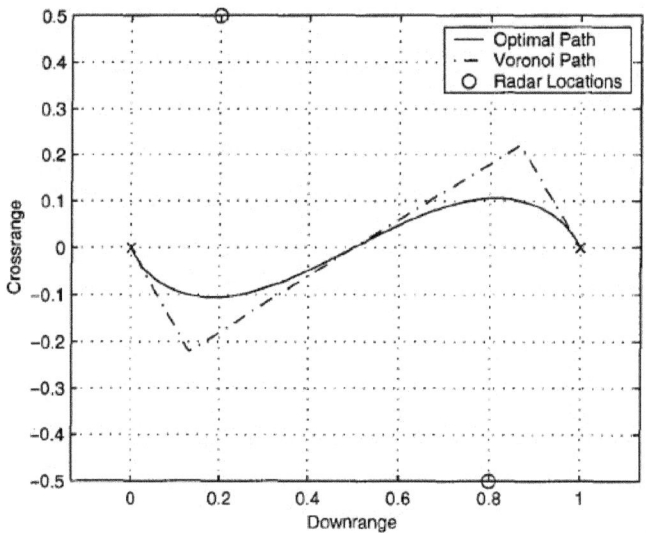

Figure A.4   Optimal Trajectory for $A = 0.6$

A-2

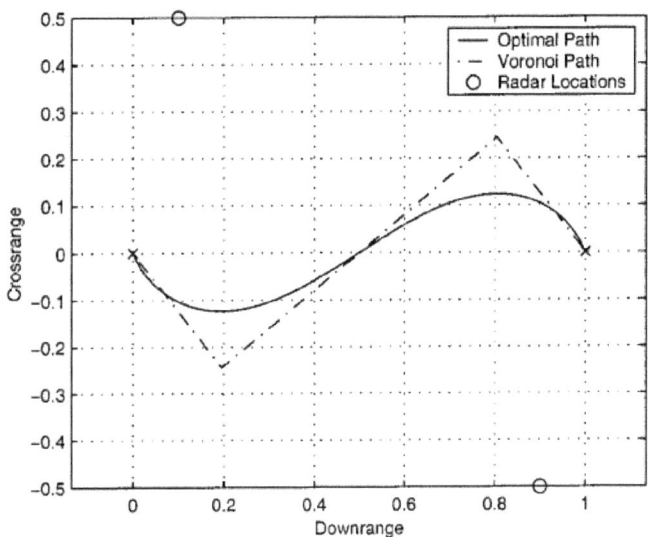

Figure A.5   Optimal Trajectory for $A = 0.8$

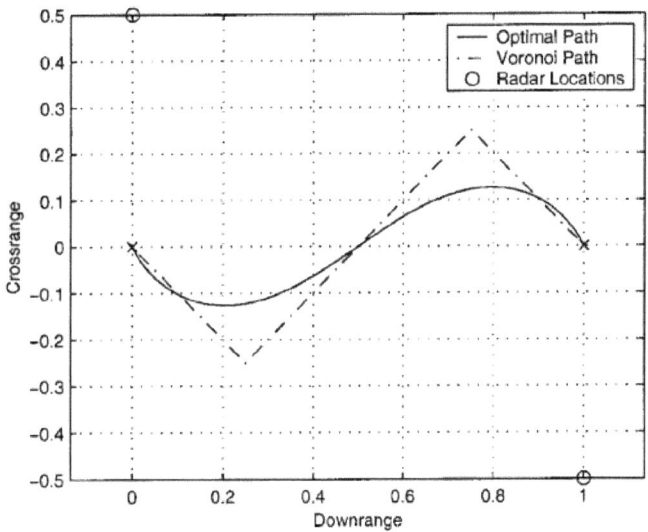

Figure A.6   Optimal Trajectory for $A = 1.0$

A.2 Scenario 2: Varying Crossrange Radar Location, $\alpha_1 = \alpha2$

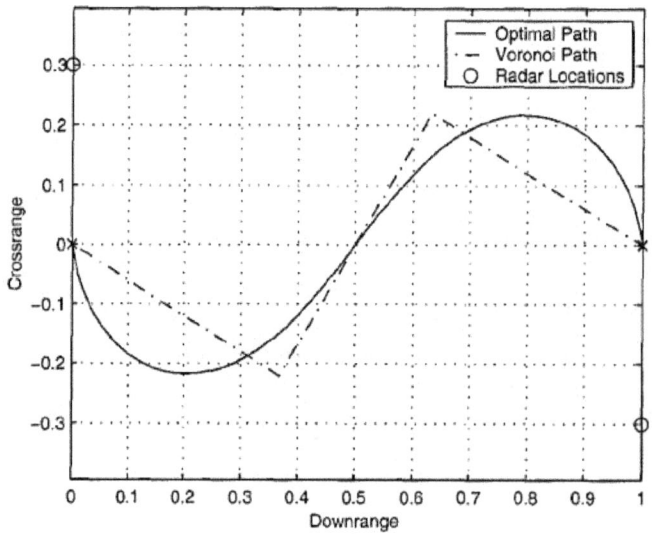

Figure A.7   Optimal Trajectory for $B = 0.6$

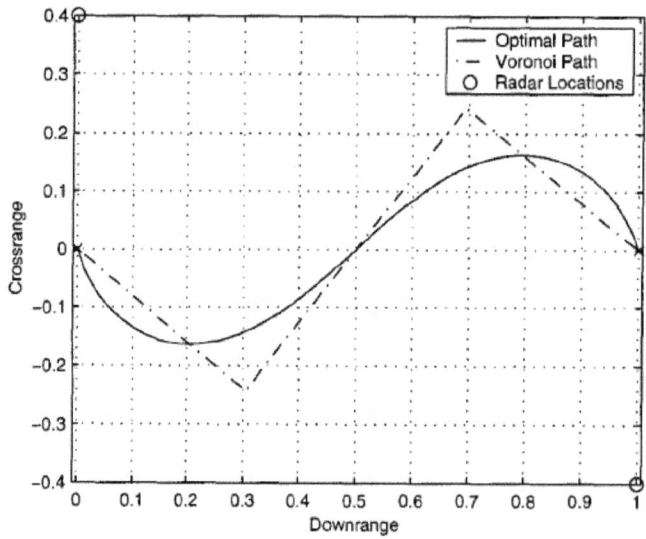

Figure A.8   Optimal Trajectory for $B = 0.8$

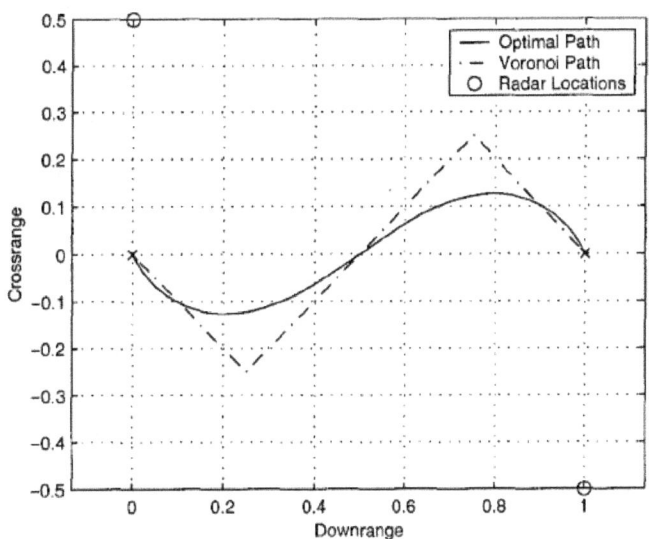

Figure A.9   Optimal Trajectory for $B = 1.0$

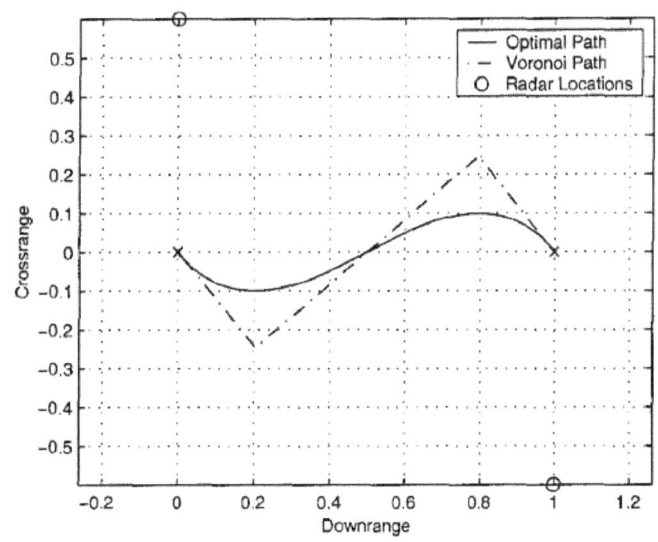

Figure A.10   Optimal Trajectory for $B = 1.2$

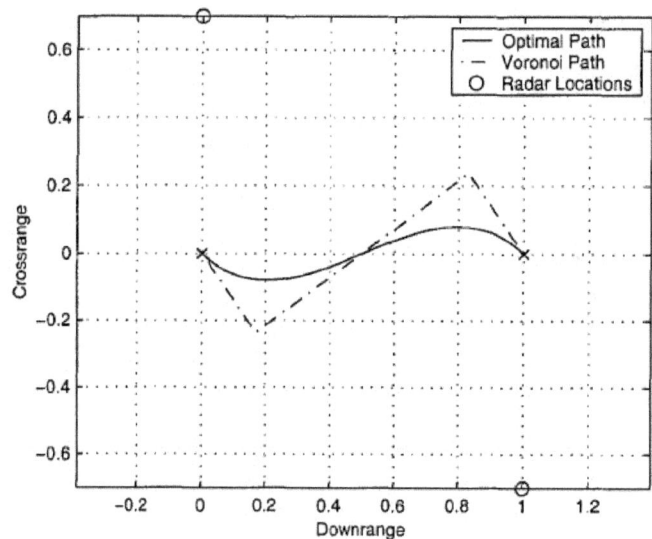

Figure A.11   Optimal Trajectory for $B = 1.4$

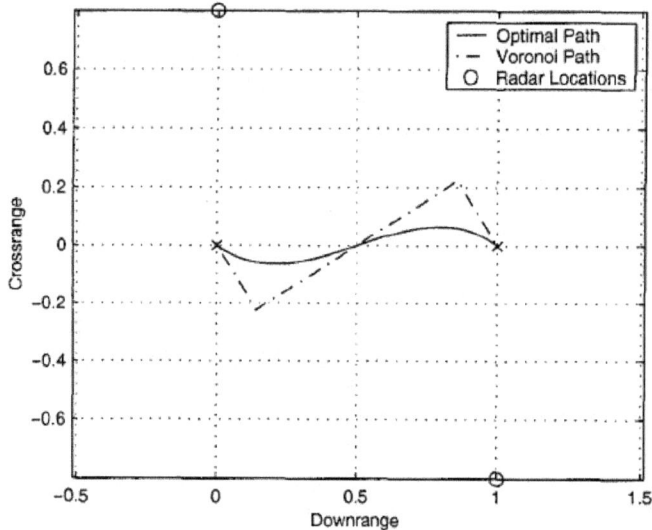

Figure A.12   Optimal Trajectory for $B = 1.6$

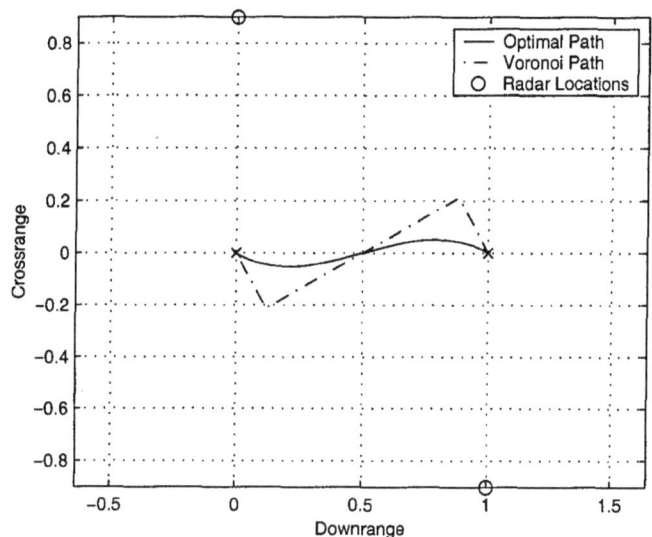

Figure A.13   Optimal Trajectory for $B = 1.8$

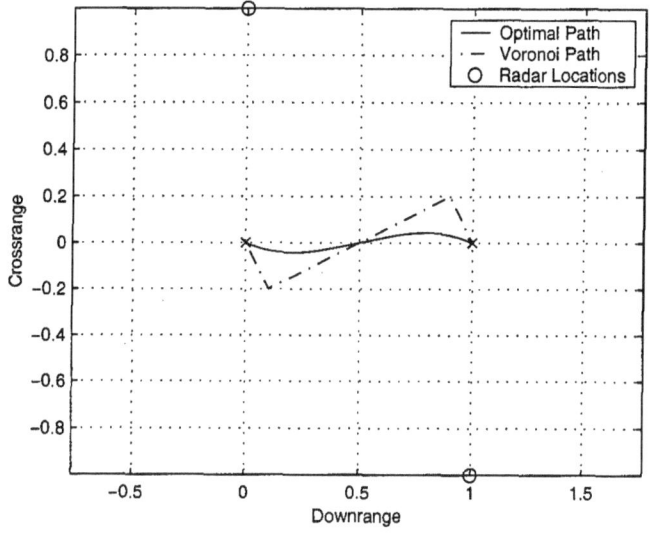

Figure A.14   Optimal Trajectory for $B = 2.0$

A-7

A.3   Scenario 3: Varying Endpoint Separation, $\alpha_1 = \alpha 2$

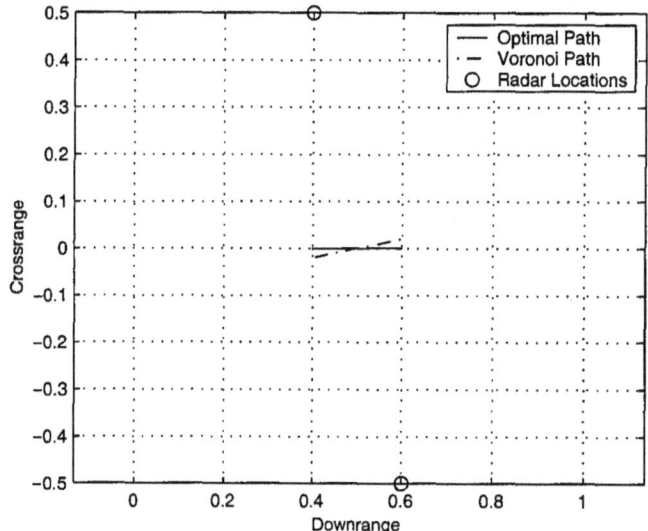

Figure A.15   Optimal Trajectory for $C = 0.2$

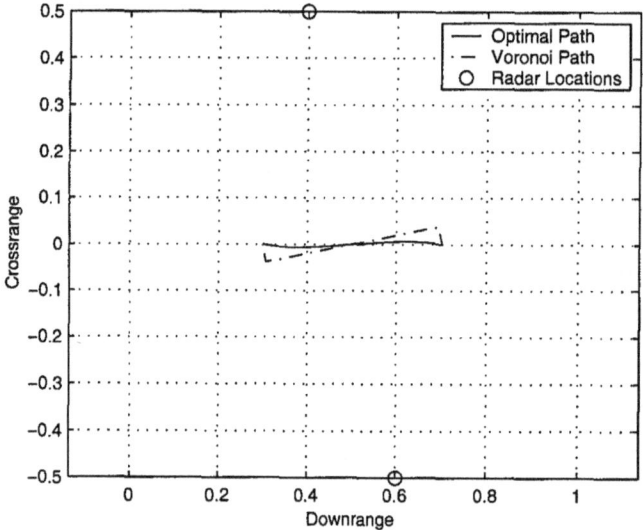

Figure A.16   Optimal Trajectory for $C = 0.4$

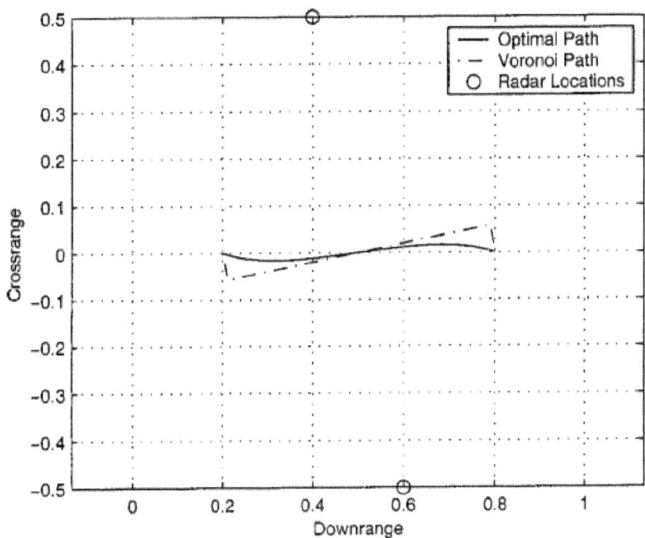

Figure A.17   Optimal Trajectory for $C = 0.6$

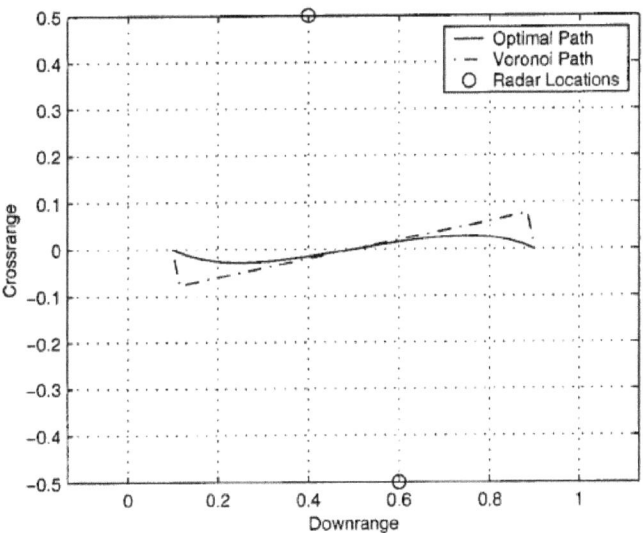

Figure A.18   Optimal Trajectory for $C = 0.8$

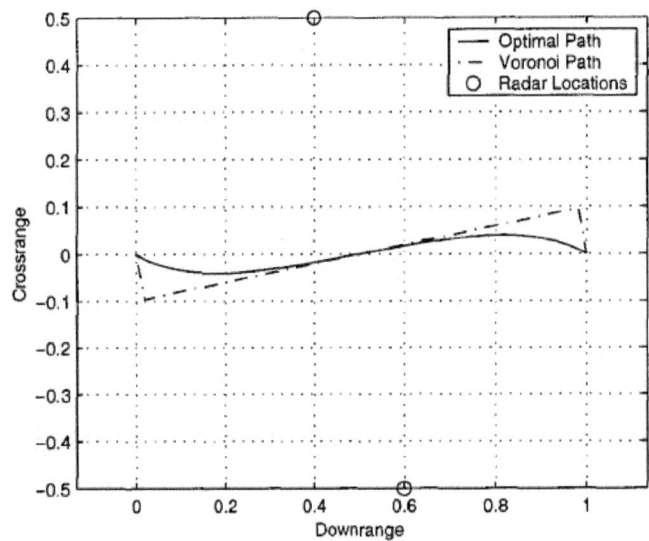

Figure A.19   Optimal Trajectory for $C = 1.0$

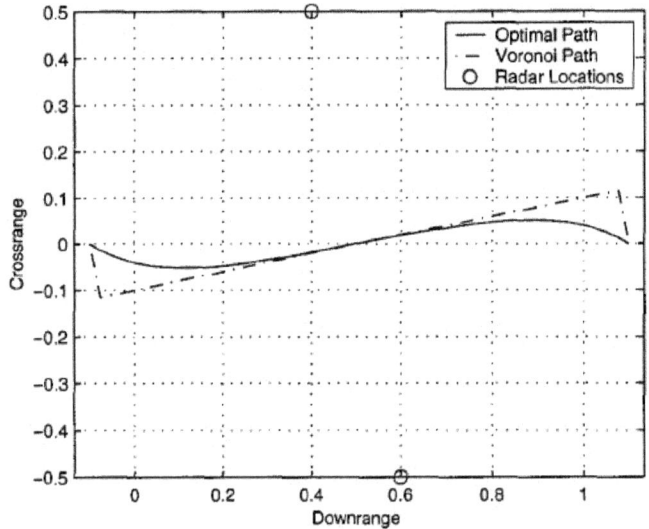

Figure A.20   Optimal Trajectory for $C = 1.2$

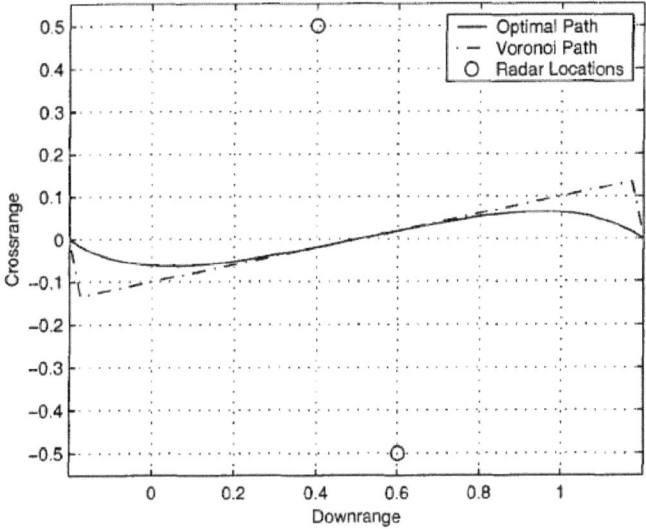

Figure A.21   Optimal Trajectory for $C = 1.4$

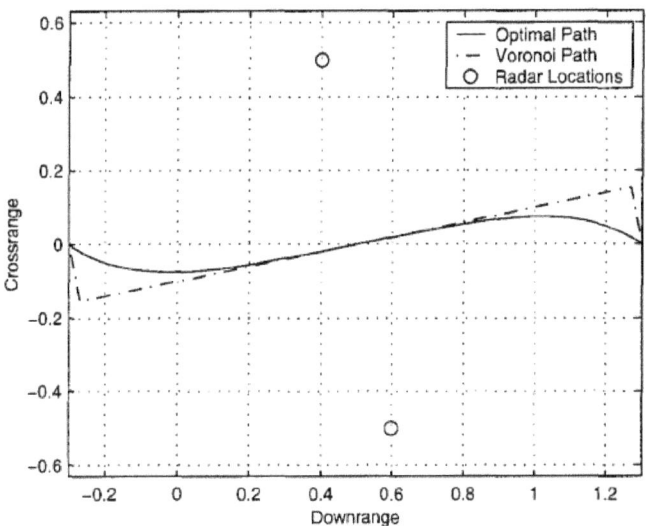

Figure A.22   Optimal Trajectory for $C = 1.6$

A-11

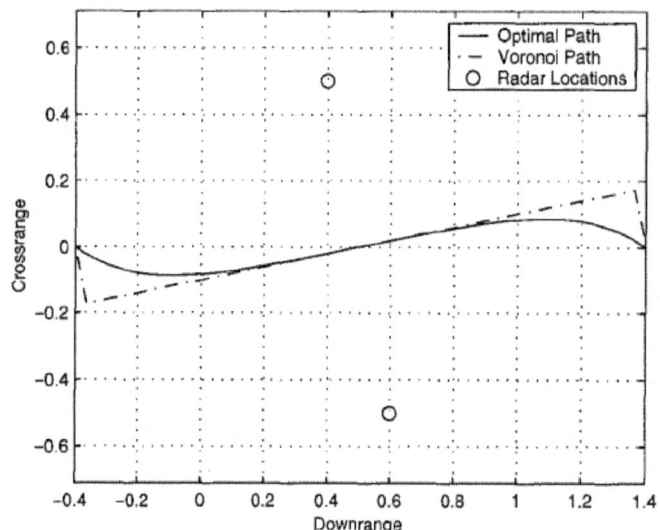

Figure A.23   Optimal Trajectory for $C = 1.8$

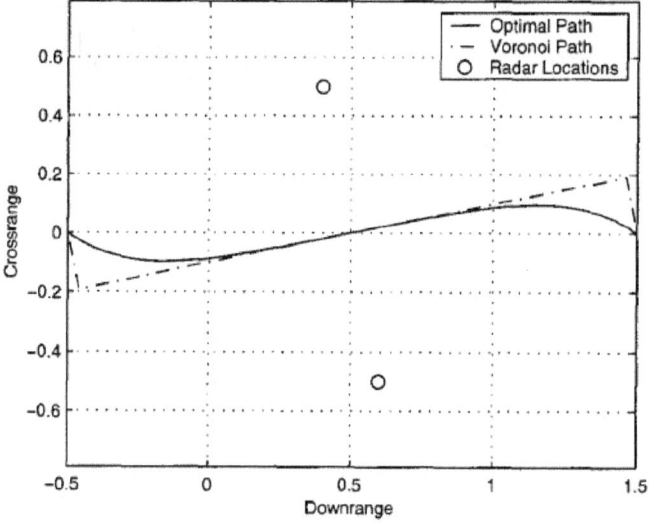

Figure A.24   Optimal Trajectory for $C = 2.0$

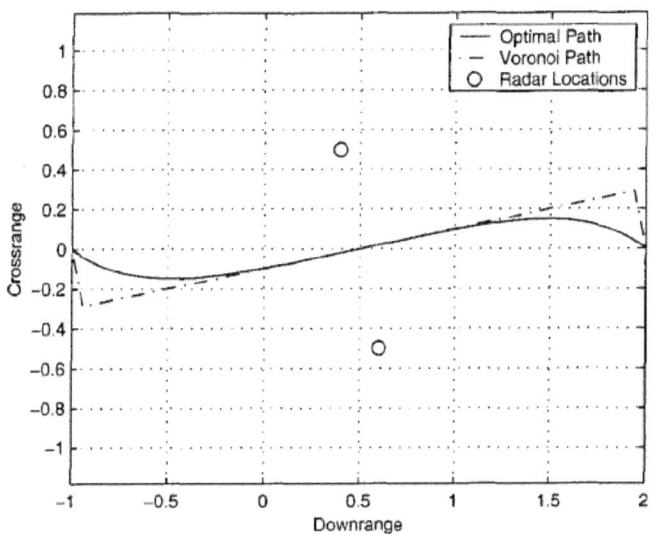

Figure A.25   Optimal Trajectory for $C = 3.0$

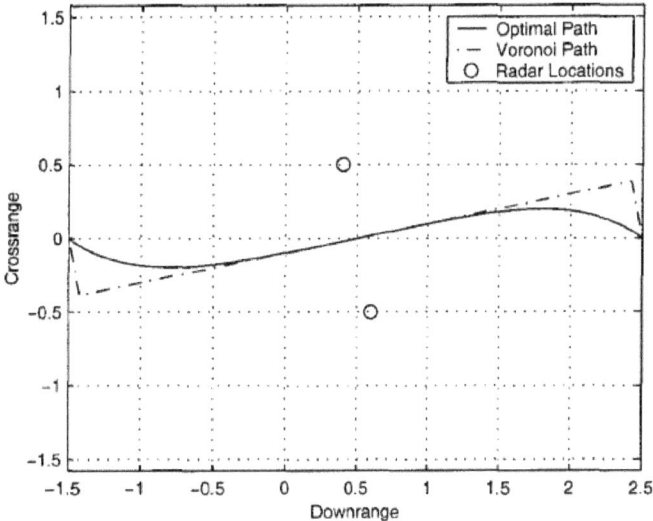

Figure A.26   Optimal Trajectory for $C = 4.0$

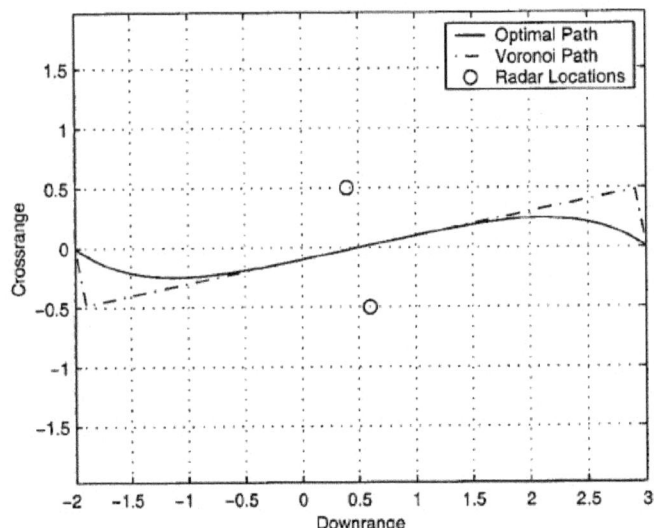

Figure A.27   Optimal Trajectory for $C = 5.0$

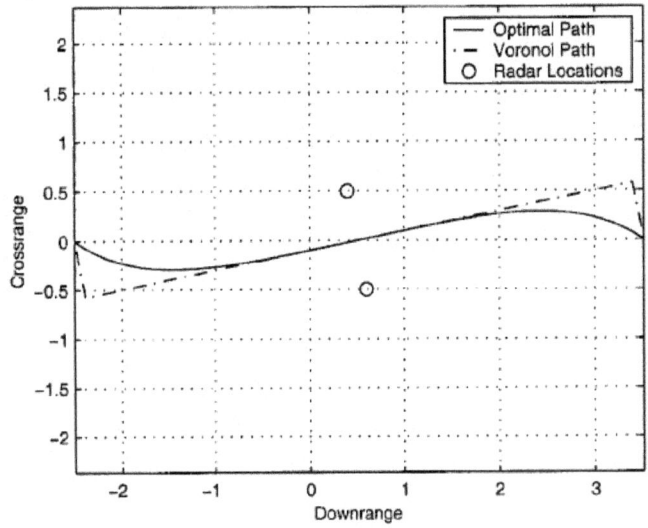

Figure A.28   Optimal Trajectory for $C = 6.0$

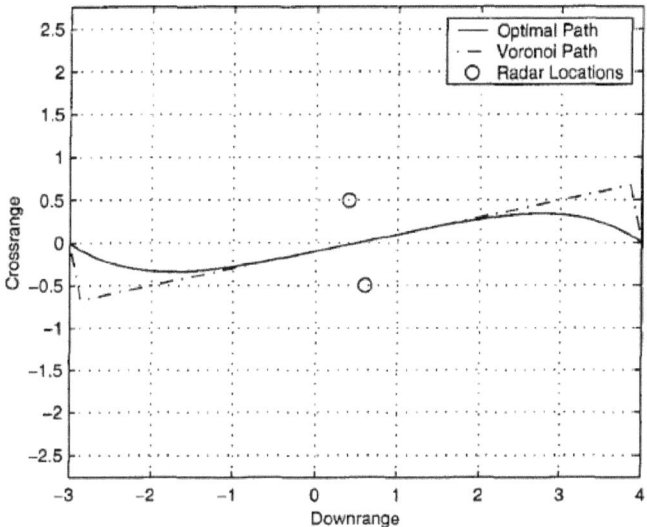

Figure A.29   Optimal Trajectory for $C = 7.0$

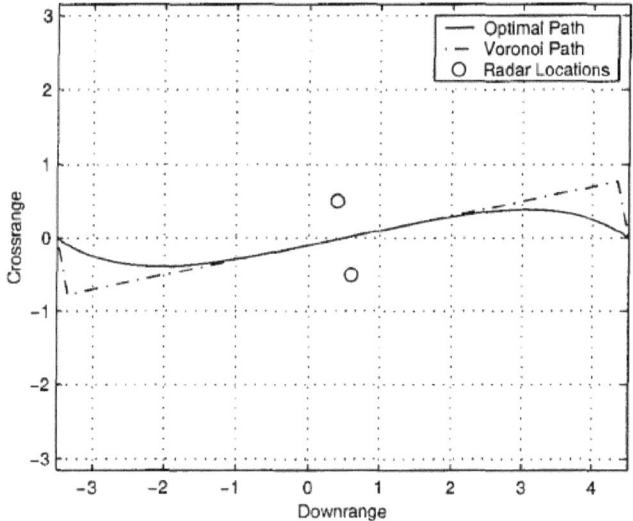

Figure A.30   Optimal Trajectory for $C = 8.0$

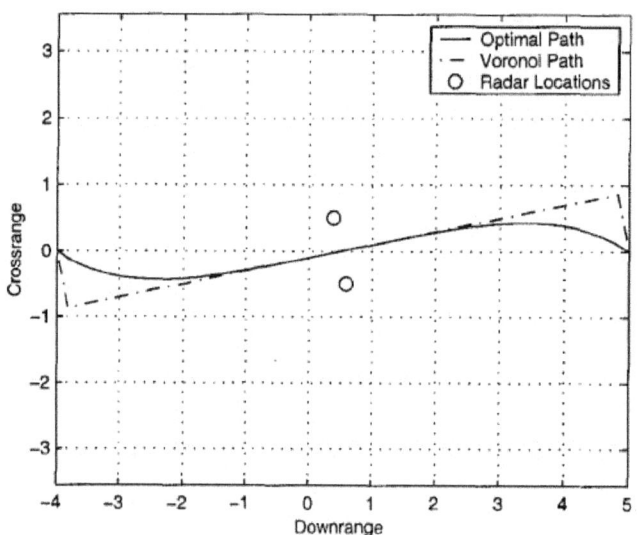

Figure A.31   Optimal Trajectory for $C = 9.0$

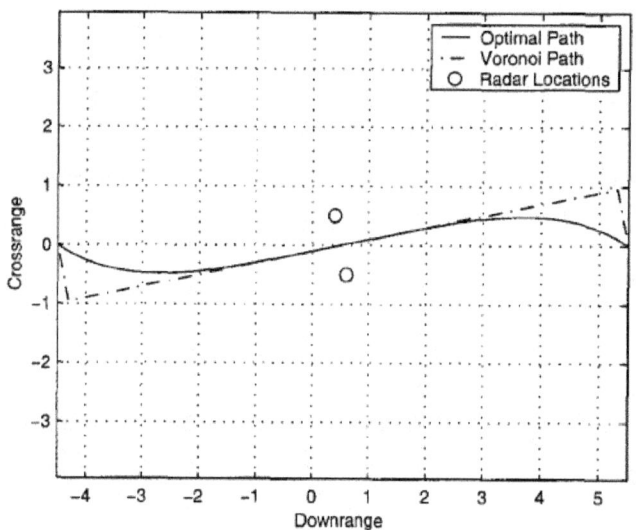

Figure A.32   Optimal Trajectory for $C = 10.0$

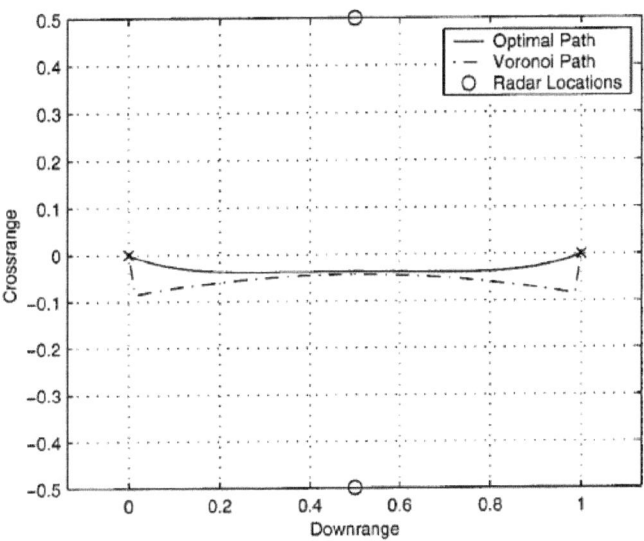

Figure A.33   Optimal Trajectory for $A = 0.0$

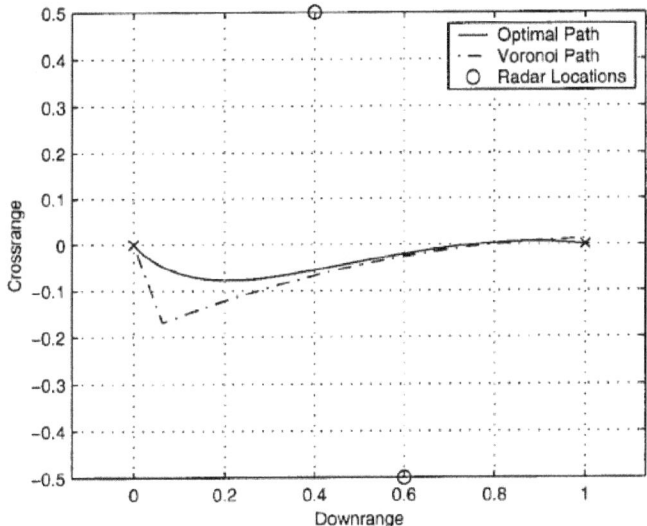

Figure A.34   Optimal Trajectory for $A = 0.2$

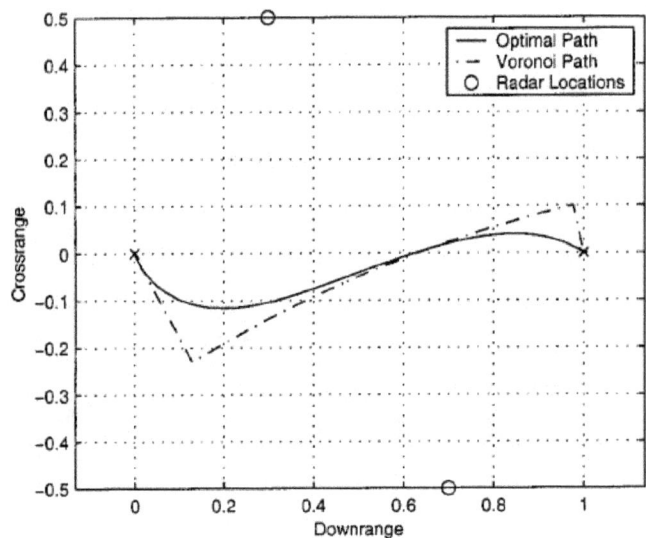

Figure A.35   Optimal Trajectory for $A = 0.4$

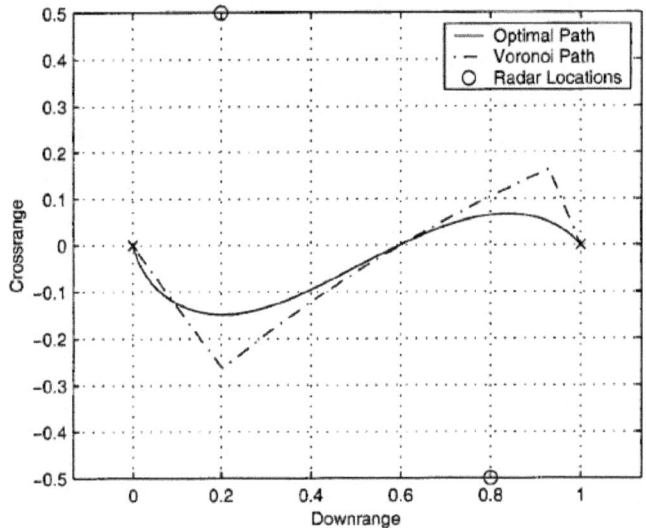

Figure A.36   Optimal Trajectory for $A = 0.6$

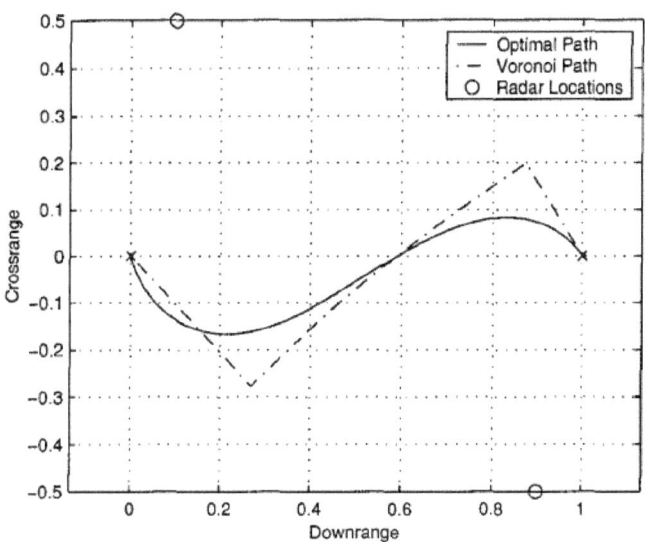

Figure A.37   Optimal Trajectory for $A = 0.8$

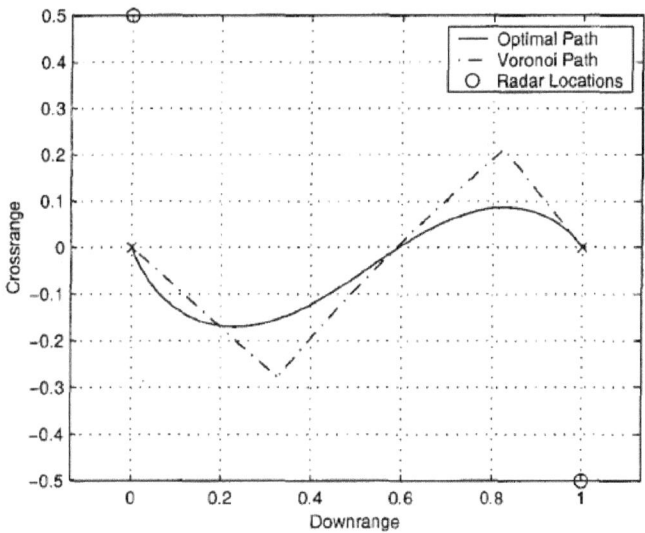

Figure A.38   Optimal Trajectory for $A = 1.0$

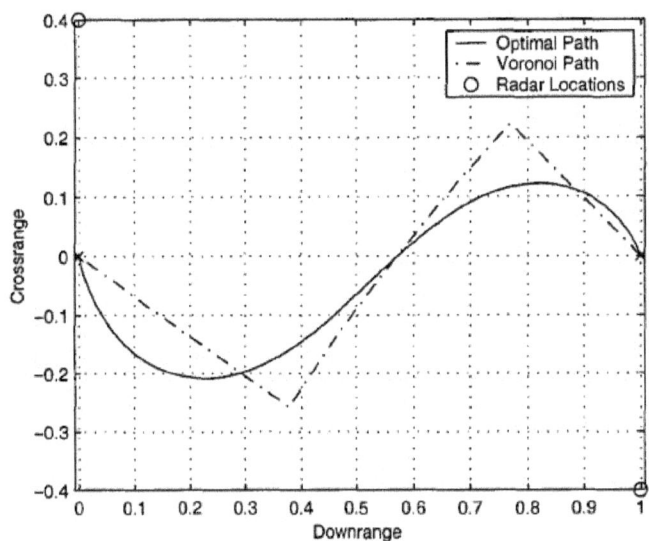

Figure A.39   Optimal Trajectory for $B = 0.8$

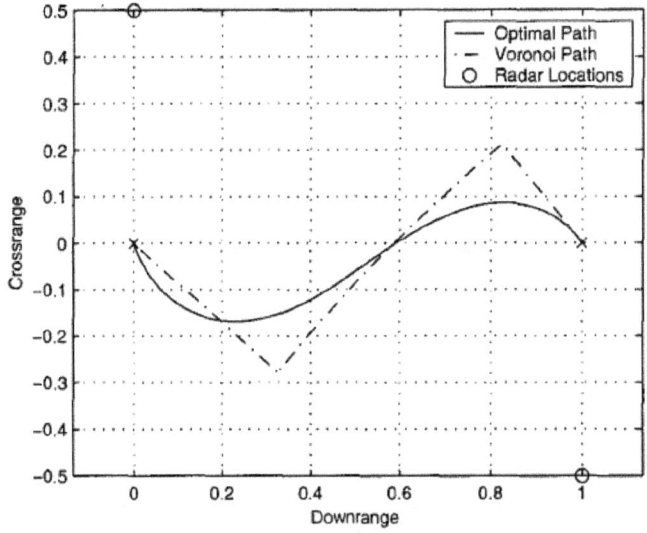

Figure A.40   Optimal Trajectory for $B = 1.0$

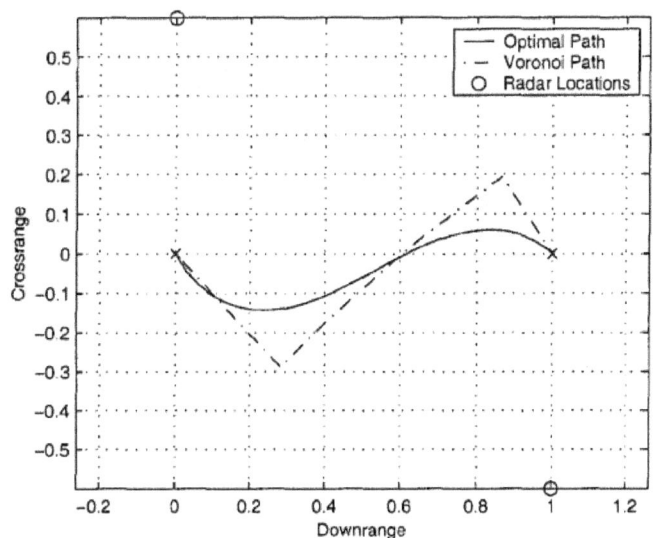

Figure A.41   Optimal Trajectory for $B = 1.2$

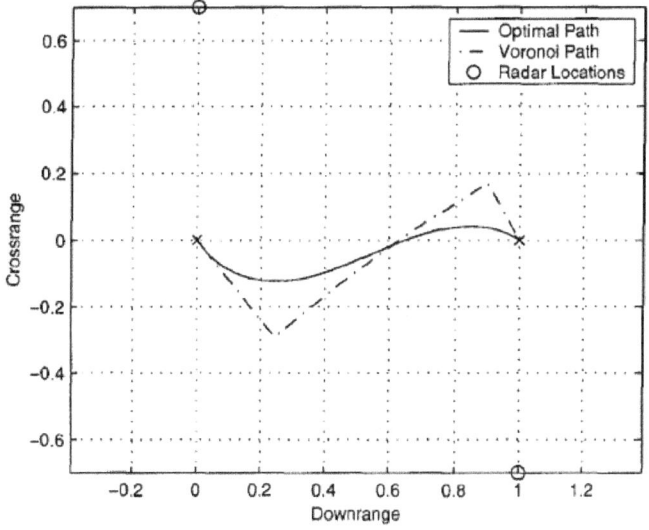

Figure A.42   Optimal Trajectory for $B = 1.4$

A-21

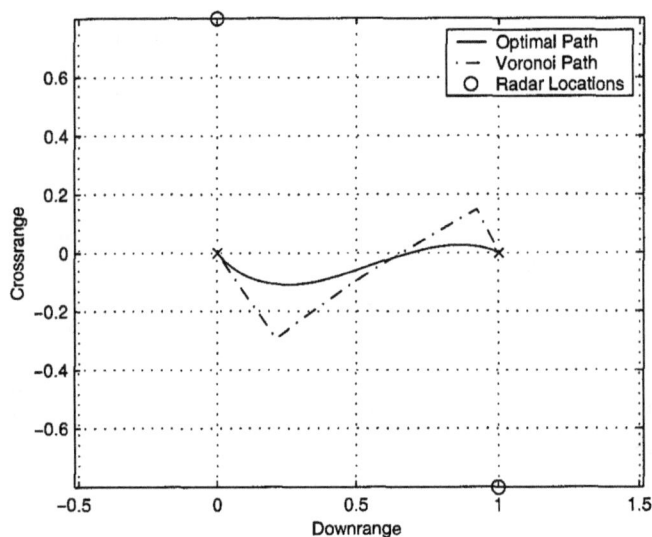

Figure A.43   Optimal Trajectory for $B = 1.6$

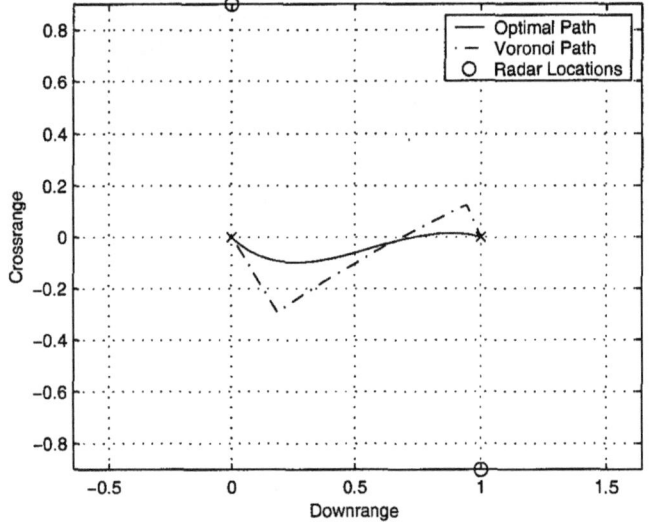

Figure A.44   Optimal Trajectory for $B = 1.8$

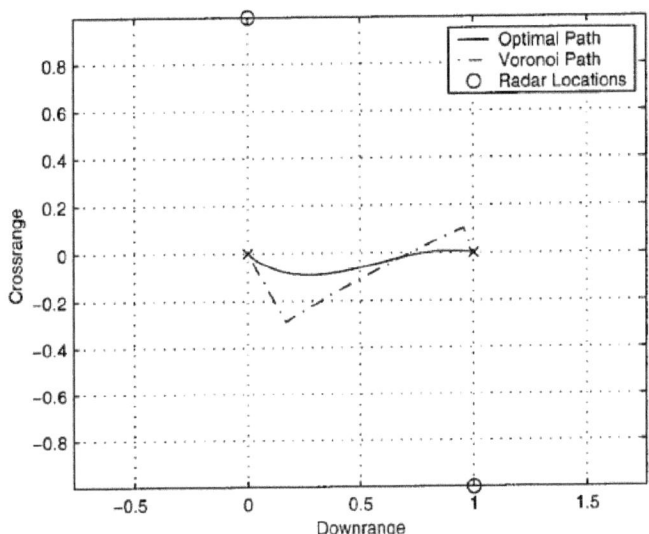

Figure A.45   Optimal Trajectory for $B = 2.0$

A.6   Scenario 3: Varying Endpoint Separation, $\alpha_1 \neq \alpha2$

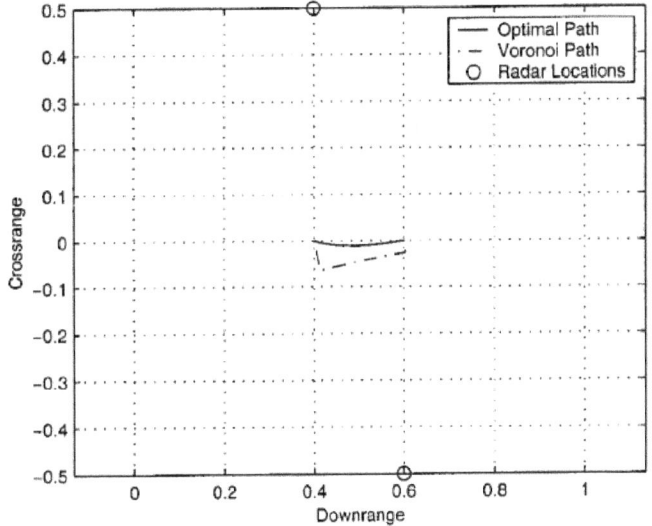

Figure A.46   Optimal Trajectory for $C = 0.2$

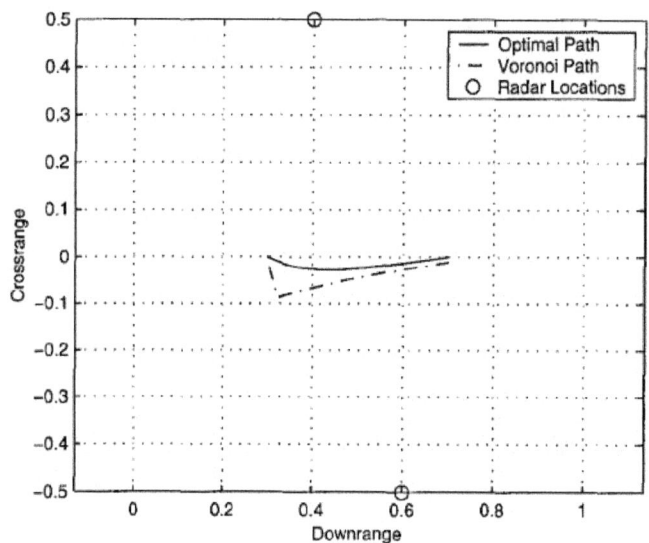

Figure A.47   Optimal Trajectory for $C = 0.4$

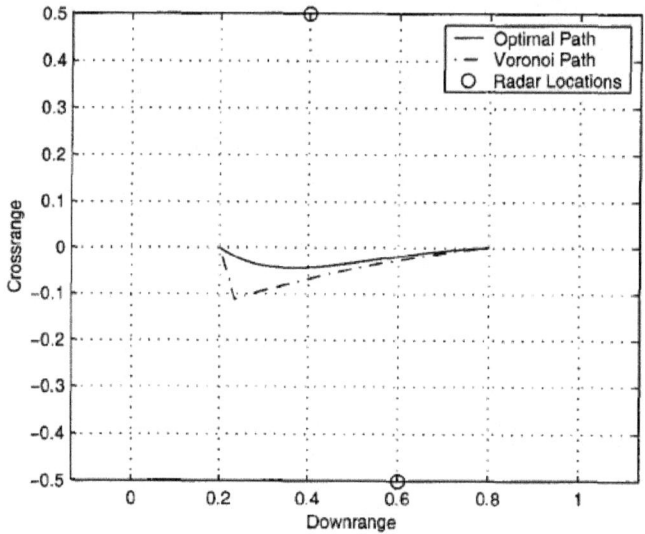

Figure A.48   Optimal Trajectory for $C = 0.6$

A-24

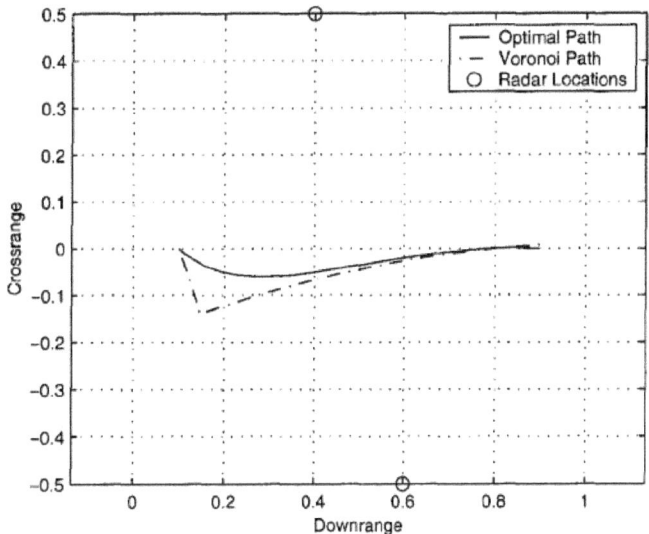

Figure A.49   Optimal Trajectory for $C = 0.8$

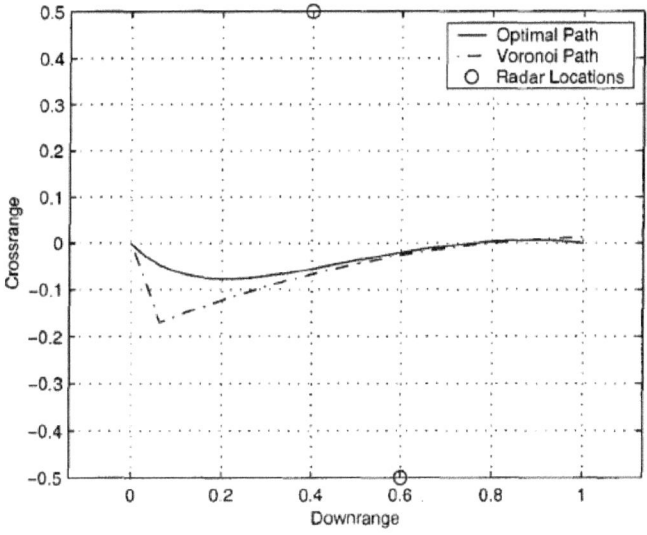

Figure A.50   Optimal Trajectory for $C = 1.0$

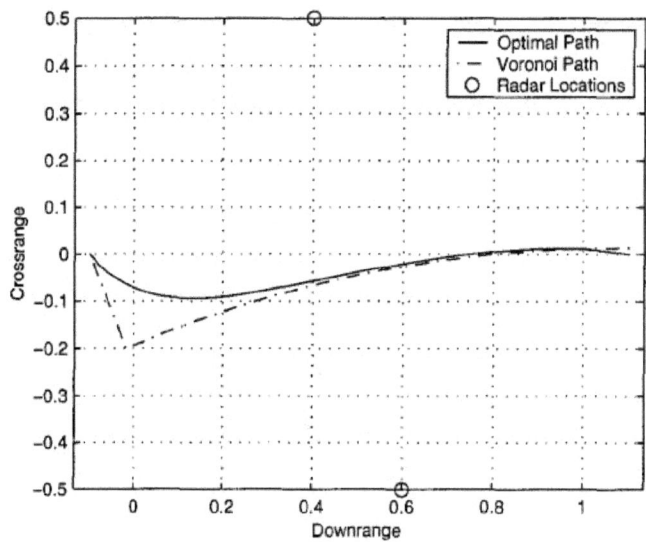

Figure A.51   Optimal Trajectory for $C = 1.2$

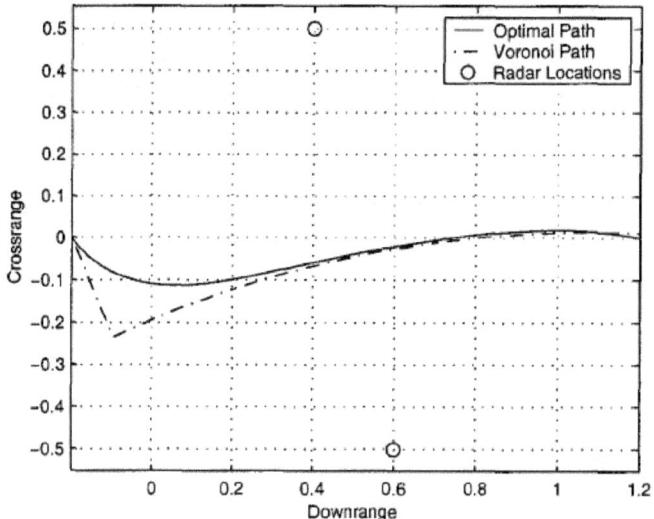

Figure A.52   Optimal Trajectory for $C = 1.4$

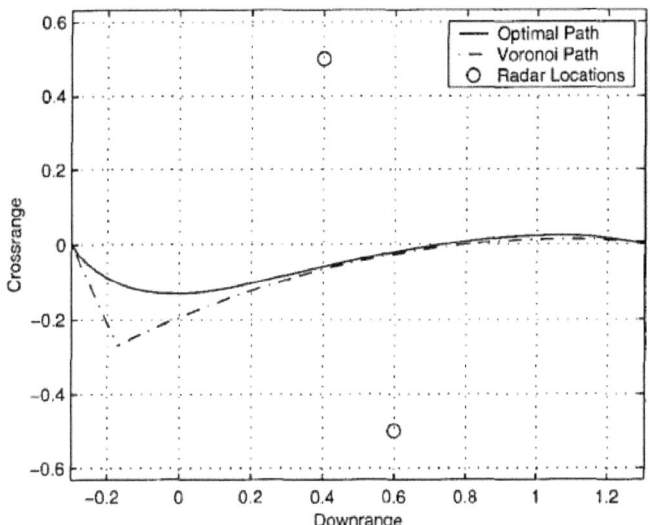

Figure A.53   Optimal Trajectory for $C = 1.6$

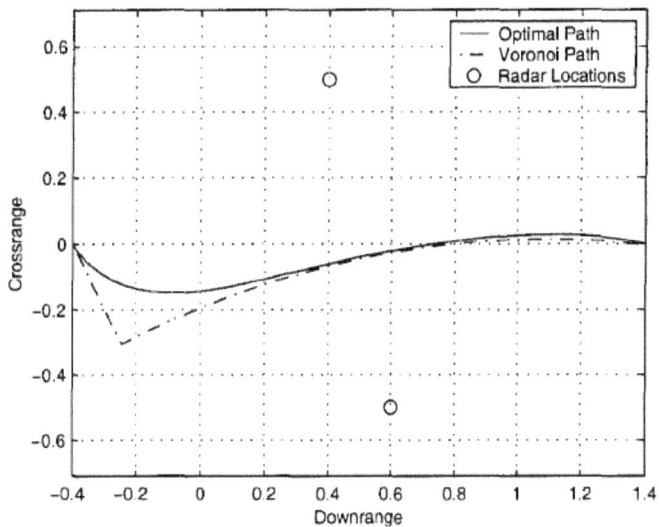

Figure A.54   Optimal Trajectory for $C = 1.8$

A-27

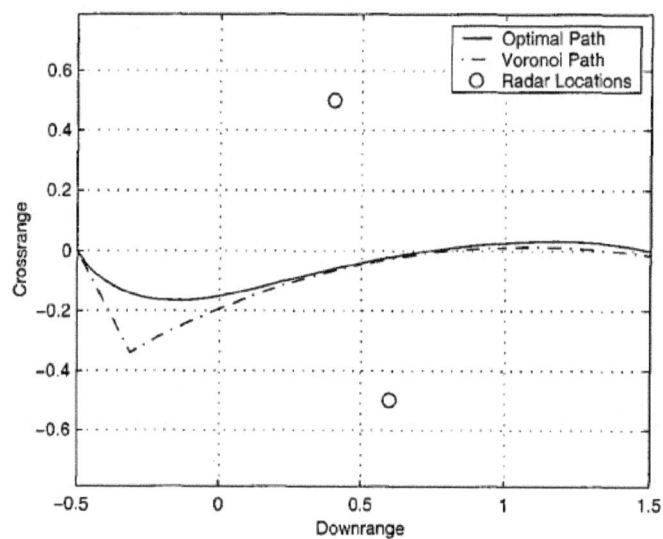

Figure A.55   Optimal Trajectory for $C = 2.0$

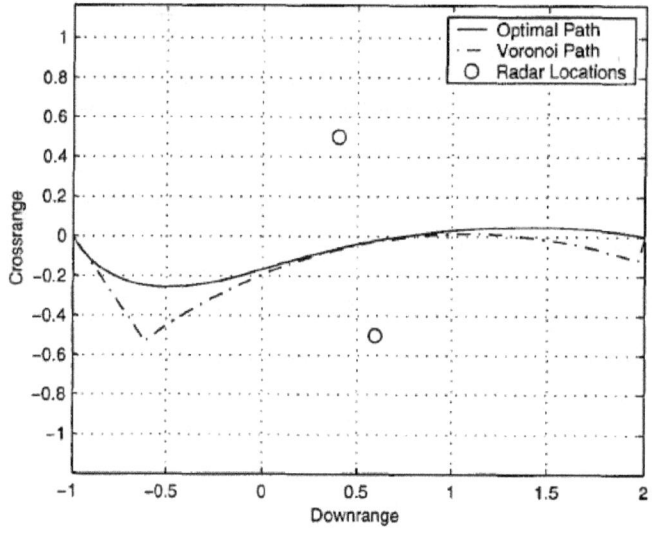

Figure A.56   Optimal Trajectory for $C = 3.0$

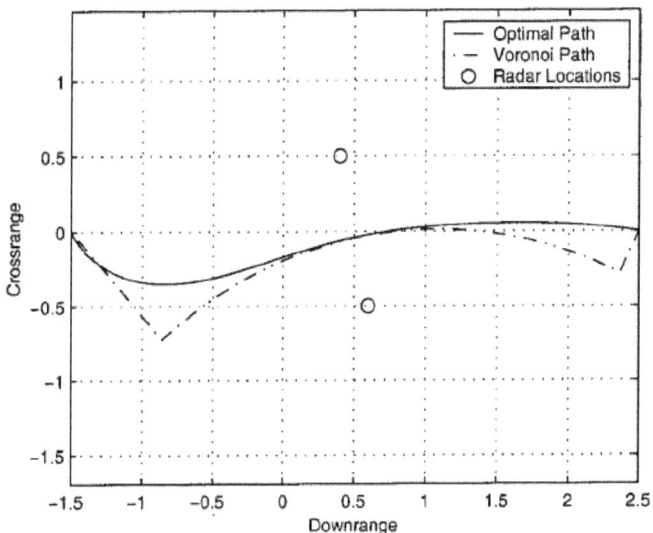

Figure A.57   Optimal Trajectory for $C = 4.0$

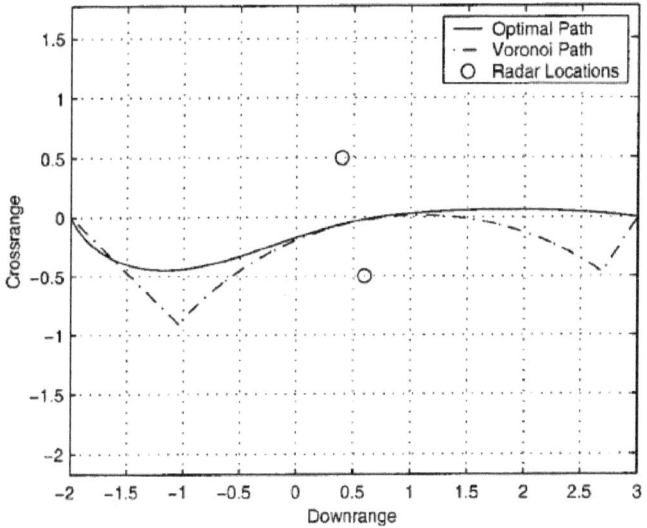

Figure A.58   Optimal Trajectory for $C = 5.0$

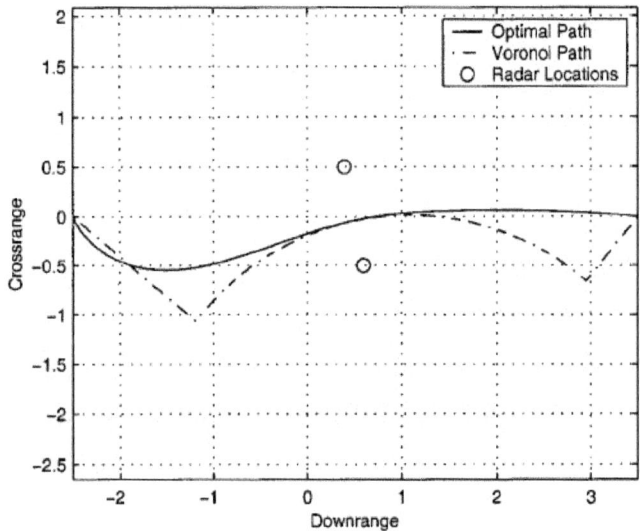

Figure A.59   Optimal Trajectory for $C = 6.0$

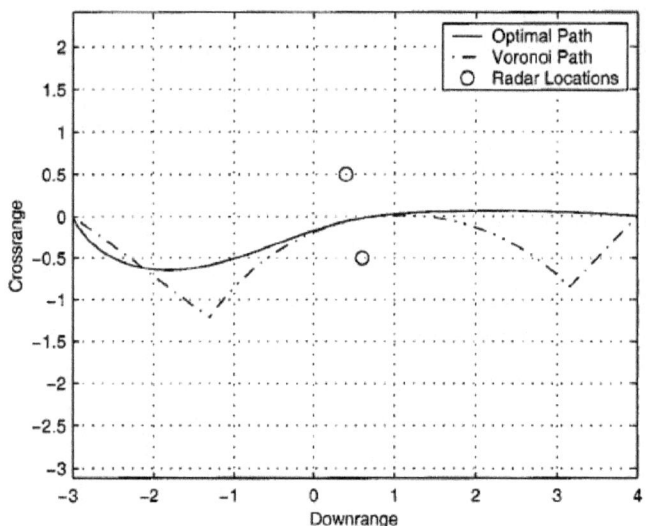

Figure A.60   Optimal Trajectory for $C = 7.0$

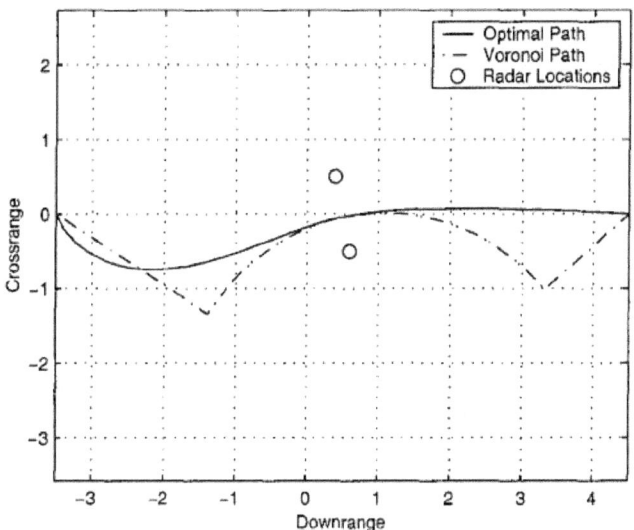

Figure A.61   Optimal Trajectory for $C = 8.0$

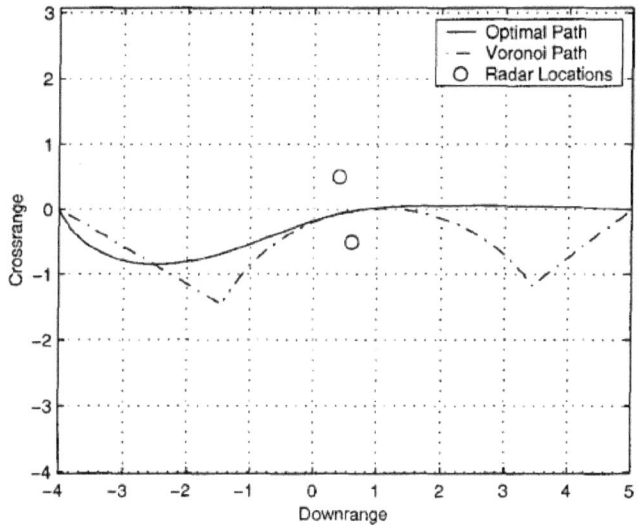

Figure A.62   Optimal Trajectory for $C = 9.0$

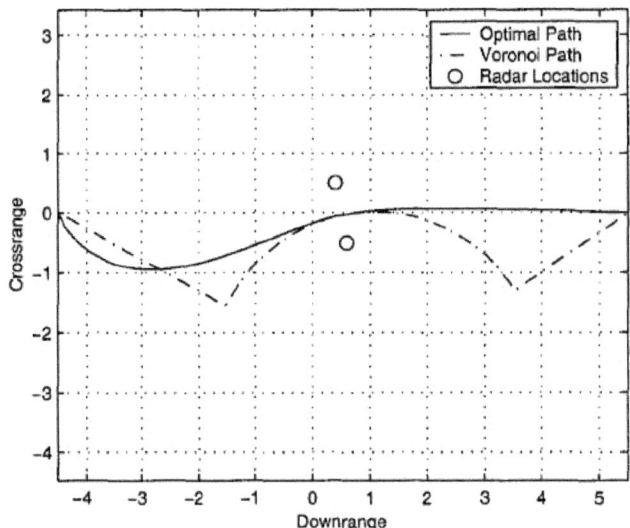

Figure A.63   Optimal Trajectory for $C = 10.0$

Bibliography

1. Apollonius, of Perga. *Conics: Books V to VII*. Ed. G. J. Toomer. New York: Springer-Verlag, 1990.

2. Borri, Marco and others. "Numerical Approach to Inverse Flight Dynamics," *Journal of Guidance, Control, and Dynamics, 20*:742–747 (July-August 1997).

3. Bortoff, Scott A. *Path-Planning For Unmanned Air Vehicles*. Unpublished Report, Wright-Patterson AFB OH: AFRL/VAAD, 1999.

4. Bortoff, Scott A. "Path Planning For UAVs." *Proceedings of the American Control Conference*. 364–368. 2000.

5. Boyle, David P. and Gregory E. Chamitoff. "Autonomous Maneuver Tracking for Self-Piloted Vehicles," *Journal of Guidance, Control, and Dynamics, 22*:58–67 (January-February 1999).

6. Bryson, Arthur E., Jr. *Dynamic Optimization*. Menlo Park CA: Addison-Wesley Longman, 1999.

7. Coleman, Thomas and others. *Optimization Toolbox For Use with MATLAB, User's Guide Version 2*. Natick MA: The Mathworks, Inc., 1999.

8. Fox, Charles. *An Introduction to The Calculus of Variations*. New York: Dover Publications, 1987.

9. Gao, C. and R.A. Hess. "Inverse Simulation of Large-Amplitude Aircraft Maneuvers," *Journal of Guidance, Control, and Dynamics, 16*:733–737 (July-August 1993).

10. Gelfand, I.M. and S. V. Fomin. *Calculus of Variations*. Englewood Cliffs NJ: Prentice-Hall, Inc., 1963.

11. Goldman, Jeffery A. "Path Planning Problems and Solutions." *Proceedings of the IEEE 1994 National Aerospace and Electronics Conference*. 105–108. New York: IEEE Press, 1994.

12. Hebert, Capt Jeffery M., PhD Student. Personal interviews. Air Force Institute of Technology (AETC), Wright-Patterson AFB OH, 2000-2001.

13. Hershey, John E. and others. "Strategic Route Planning and Sensor Fusion," *IEEE Transactions on Aerospace and Electronic Systems, 29*:1357–1359 (October 1993).

14. Hess, R.A. and others. "Generalized Techniques for Inverse Simulation Applied to Aircraft Maneuvers," *Journal of Guidance, Control, and Dynamics, 14*:920–926 (September-October 1991).

15. Hull, David G. "Conversion of Optimal Control Problems into Parameter Optimization Problems," *Journal of Guidance, Control, and Dynamics, 20*:57–60 (January-February 1997).

16. Jacques, David R. and others. "A MATLAB Toolbox for Fixed-Order, Mixed-Norm Control Synthesis," *IEEE Control Systems Magazine, 15*:36–44 (October 1996).

17. Jung, Y.C. and R.A. Hess. "Precise Flight-Path Control Using a Predictive Algorithm," *Journal of Guidance, Control, and Dynamics, 14*:936–942 (September-October 1991).

18. Krozel, Jimmy and Dominick Andrisani II. "Navigation Path Planning for Autonomous Aircraft: Voronoi Diagram Approach," *Journal of Guidance, Control, and Dynamics, 13*:1152–1154 (November-December 1990).

19. Lewis, Frank L. and Vassilis L. Syrmos. *Optimal Control*. New York: Wiley Interscience, 1995.

20. Lietmann, George (ed.). *Optimization Techniques*. London, England: Academic Press, Inc., 1963.

21. Lin, C.F. and L.L. Tsai. "Analytical Solution of Optimal Trajectory-Shaping Guidance," *Journal of Guidance, Control, and Dynamics, 10*:61–66 (January-February 1987).

22. McFarland, Michael B. and others. "Motion Planning for Reduced Observability of Autonomous Aerial Vehicles." *Proceedings of the 1999 IEEE Conference on Control Applications*. 231–235. IEEE Press, 1999.

23. Meng, A. "Flight Path Planning Under Uncertainty For Robotic Air Vehicles." *IEEE National Aerospace and Electronics Conference*. 359–366. IEEE Press, May 1987.

24. Min, Yi and others. "3D Route Planning Using Genetic Algorithm." *Proceedings of the SPIE - The International Society for Optical Engineering Vol 3545*. 92–95. SPIE-International Society for Optical Engineering, 1998.

25. Okabe, Atsuyuki and others. *Spatial Tessellations: Concepts and Applications of Voronoi Diagrams*. Chichester, England: Wiley & Sons, 1992.

26. Pachter, Meir and others. "Minimizing Radar Exposure in Air Vehicle Path Planning." *Proceedings of the 41st Israel Annual Conference on Aerospace Sciences*. Feb 2001.

27. Pellazar, Miles B. "Vehicle Route Planning with Constraints Using Genetic Algorithms." *Proceedings of the IEEE 1994 National Aerospace and Electronics Conference*. 111–118. New York: IEEE Press, 1994.

28. Pendelton, Capt Ryan R. *Use of Unusual Aircraft Orientations To Generate Low Observable Routes*. MS thesis, AFIT/GAE/ENY/00M-09, Air Force Institute of Technology, Wright-Patterson AFB OH, March 2000.

29. Press, William H. and others. *Numerical Recipes in C*. New York: Cambridge University Press, 1992.

30. Psiaki, Mark L. and Kihong Park. "Trajectory Optimization for Real-Time Guidance; Part 1, Time-Varying LQR on a Parallel Processor." *Proceedings of the American Control Conference*. 248–253. Piscataway NJ: IEEE Press, 1990.

31. Rao, Niranjan S. and others. "Singular Perturbation Based Aircraft Trajectory Optimization for Threat Avoidance." *Proceedings of the American Control Conference*. 3114–3115. IEEE Press, 1990.

32. Roberts, Sanford M. and Jerome S. Shipman. *Two-Point Boundary Value Problems: Shooting Methods*. New York: American Elsevier, 1972.

33. Selvestrel, Mario C. and Simon Goss. "Optimal Flight Paths for Aircraft: A Little Knowledge Goes a Long Way." *IEA/AIE Proceedings of the Eighth International Conference*. 687–695. Gordon and Breach Science Publishers, 1995.

34. Sentoh, Etsuroh and Arthur E. Bryson. "Inverse and Optimal Control for Desired Outputs," *Journal of Guidance, Control, and Dynamics*, *15*:687–691 (May-June 1992).

35. Shapira, Ilana and Joseph Z. Ben-Asher. "Near-Optimal Horizontal Trajectories for Autonomous Air Vehicles," *Journal of Guidance, Control, and Dynamics*, *20*:735–741 (July-August 1997).

36. Shiller, Zvi and J.C. Chen. "Optimal Motion Planning of Autonomous Vehicles in Three Dimensional Terrains." *Proceedings of the 1990 IEEE International Conference on Robotics and Automation*. 198–203. Los Alamitos CA: IEEE Computer Society Press, 1990.

37. Skolnik, Merrill I. *An Introduction to Radar Systems*. New York: McGraw-Hill, Inc., 1980.

38. Vian, John L. and John R. Moore. "Trajectory Optimization with Risk Minimization for Military Aircraft," *Journal of Guidance, Control, and Dynamics*, *12*:311–317 (May-June 1989).

39. Visual Numerics, Inc. *IMSL Math Library, 1*. Visual Numerics, Inc., 1997.

40. Yakimenko, Oleg A. "Direct Method for Rapid Prototyping of Near-Optimal Aircraft Trajectories," *Journal of Guidance, Control, and Dynamics*, *23*:865–875 (September-October 2000).

www.ingramcontent.com/pod-product-compliance
Lightning Source LLC
Chambersburg PA
CBHW080708190526

45169CB00006B/2293